Julian-Banner Mining District

A Road to Gold

Photo of gold discovered by local
San Diego geologist and father in
Plumas County, California, circa 1995,
Richard Voss and Bryan Voss.
Photo provided courtesy of Bryan Voss.

Julian-Banner Mining District: A Road to Gold

 First edition 2017, fourth printing 2022. Edited by Christopher D. Livesey

Cover photo Adobe Stock (www.stockadobe.com), modified by Swanson Photography http://www.rswansonphotography.com
Book design and composition by Michael Schrauzer
Printed in United States of America

San Diego Geological Society
3130 North Evergreen Street, San Diego, California 92110
www.sandiegogeologicalsociety.org

Sunbelt Publications
1250 Fayette Street, El Cajon, California 92020
(619) 258–4911, fax: (619) 258–4916
www.sunbeltpublications.com

25 24 23 22 7 6 5 4

Library of Congress Cataloging-in-Publication Data

Names: Livesey, Christopher D., 1981- editor. | San Diego Association of Geologists.
Title: Julian-Banner District : a road to gold : guidebook for a field trip of the San Diego Association of Geologists, October 15-16, 2017 / editor, Christopher D. Livesey.
Other titles: Julian-Banner Mining District : a road to gold
Description: San Diego, California : San Diego Geological Society, [2017] | Includes bibliographical references.
Identifiers: LCCN 2018000659 | ISBN 9781941384404 (softcover : alk. paper)
Subjects: LCSH: Geology--California, Southern--Guidebooks. | California, Southern--Guidebooks.
Classification: LCC QE90.S65 J85 2017 | DDC 557.94/985--dc23
LC record available at https://lccn.loc.gov/2018000659

The San Diego Geological Society, Inc., gratefully acknowledges the generous contributions of our sponsors and guidebook contributors to the publication of this book.

Front and Back Cover Photograph: "Unknown Prospector in Unknown Location," photograph modified and designed by Swanson Photography
Inside Front Cover Photograph: Courtesy of Bryan Voss

PLEASE READ: This book describes driving and hiking situations and other potentially hazardous activities that are the sole responsibility of the reader. Exploring geological features in mixed terrain from rural highways, remote mountain wilderness, to urban environments is not without risk. All who engage should be aware of their own capabilities, be knowledgeable, prepared, and alert. When exploring such environments, please abide by local laws, respect regional customs and private property, and don't forget about common sense. The conditions, access, and features described in this book can and do change with time.

JULIAN-BANNER MINING DISTRICT

A Road to GOLD

GUIDEBOOK FOR A FIELD TRIP

of the

San Diego Association of Geologists

OCTOBER 15–16, 2017

Christopher D. Livesey, *Editor*

SAN DIEGO ASSOCIATION OF GEOLOGISTS

Distributed by

SUNBELT PUBLICATIONS

Photo of unknown mine, courtesy of David Lewis of the Julian Pioneer Museum.

TABLE OF CONTENTS

Looking into Eagle Mine adit .
Photo courtesy of Eagle Mining Company, www.theeaglemining.com.

Lode in Eagle Mine adit.

Photo courtesy of Eagle Mining Company, www.theeaglemining.com.

PROLOGUE

I WANT TO TAKE THE TIME TO THANK YOU, THE READER, for the trip you are about to traverse. Creating the fieldtrip and guidebook has been, in itself, a long journey. I hope you enjoy the trip as much as I had.

I grew up in the county suburbs of Los Angeles and Orange California, but traveled yearly into the wild mountains for backpacking trips. In my young adult life, exploration atop the mountains waned, but my past un-doubtfully shaped my future passion and career in geology.

I never realized how surreal and tranquil the mountains are, until I return to the arterial streets of a condensed population. The city in itself is an eco system that breathes in and exhales cars at a nearly continuous period. It should be a requirement to recess into nature to understand the fundamentals of life.

Geology is a passion for some, a profession for others, and a journey for many. The subject of geology brings a diverse group of people together under the crisp blue skies and infinite night skies. Case in point, during my long education career as an undergraduate, I noticed geology intrigued many students, to a point to change their declaration of major to geology. Although I cannot compare this experience to other majors, I never heard a student exclaim in any other class (English, math, physics, chemistry, or biology): "Wow, this is really interesting I need to change my major to… (insert non-geology major)"

I have composed this guidebook to appeal to the quick reader and the leisure reader.

For the quick reader, I have provided a summary of the road log herein under Part I. This will provide you with a quick path to the points of interest along State Highway 78.

Part II presents a more detailed road log orientated for the more intense traveler.

Part III provides papers that were specifically written to supplement some of the roadside stops and, are also, noted in the road log as supplemental reading material that will discuss in more detail the point of interest. Other papers provide some enlightenment into other mining aspects of the region, specifically, gem quality, pegmatite minerals and a look at Julian beyond the era of the gold rush.

I hope you will find a well cohesive guidebook that will take you as deep into the experience as you would like. In an effort to generally summarize the geology of Ramona and Julian that can be associated with the mineralization of gold: The trip will take you to the heart of a pluton (Stop no. 1 Ramona Plutonic Complex and Stop No. 3), the result of the subducting Faralon tectonic plate beneath the western United States. As the Faralon plate, laden with gold bearing sediments subducted, metamorphism transformed the sediment into Julian Schist. Later events thrusted up the Julian Schist, which allowed the gold to eventually be found by gold thirsty prospectors; initially in the natural creek drainages (Stop no. 4 and 5) and later in the lode bearing hillside of Julian and Banner Grade (Stops no. 8, 9, and 10).

Enjoy.

Christopher D. Livesey
Vice President of SDAG 2017
Field Trip Organizer

ACKNOWLEDGMENTS

THERE IS NO DOUBT THAT PATIENCE AND PERSERVERence (as my father always says) are the two most crucial skills required when attempting to organize a fieldtrip, especially a successful one! Close seconds would be collaboration and cooperation, and there are many people that should be recognized for their hard work and support in this endeavor. In no particular order and in hopes of not excluding a singularity: My wife has been especially considerate, as I spent several weekdays venturing prospects in gold while she stayed home with our infant and tending our own twinning gemstones, ironically to be delivered within a week of the planned fieldtrip. My wife deserves a lot of credit for this act of understanding. I would also like to thank in no particular order: Pat Abbott, Bill Elliot, Mike Thacker, Monte Marshall, Monte and Diane Murbach, Martin Learn, Lowell Lindsay, Bob Stroh, Adam Avakian, Ken Haase, Jen Morton, Bryan Voss, the operation staff at Hollow Creek Ranch, Toni and Mike Menghini, Carole and Mike Ziegler for sponsoring the Menghini wine during Saturday night, The San Diego Historical Society, Gemological Institute of America, JR Morgan, Dave Kimbrough, Marty Grove, Mike Jensen, Mary Murphy, Carolyn Glockhoff, and Chuck Houser.

Christopher D. Livesey

FIELD TRIP COMMITTEE

Christopher D. Livesey — Monte Murbach
Lowell Lindsay — Ken Haase
Jennifer Morton — Adam Avakian
Diane Murbach — Bob Stroh

FIELD TRIP LEADERS AND GUIDEBOOK CONTRIBUTORS

Pat Abbott — Dave Kimbrough
Bill Elliot — Marty Grove
Christopher D. Livesey — Brendan M. Laurs
Monte Murbach — Kimberly Knox
Martin Learn — Mike Thacker
JR Morgan — Monte Marshall
Kathryn A. Jordan

IN MEMORIAM

The crystal sky still pricked with stars
        breaks on my consciousness.
The air is quiet as if in wait
        for the first gentle fingers of wind;
And my body aches for the peace
        that these signs should bring.
But that part of my heart and mind
        which was devoted to loving and caring
Has been torn from me by the knowledge
        that this earth will never see
                those footprints again.
And the morbid muse has touched me
        with alternatives which can never
                be used to affect what has been done.

The morning wind is upon me
        and the pink edge of dawn
has seeped into
the mantle of darkness
and the certainty that life
        must continue—
either here or beyond—
consoles my spirit and braces me
        for the celebration
of life and living around me.

I hide my sorrow
        so that time can heal
the ulcer in my soul.
So my burning agony
        will not blind my loved ones
                to the joys of living;
I partake of life;
        even as the disciples broke bread,
with love and remembrance
        of a time-freed soul.

James R. Evans 1976
(Edited KJE)

DEDICATIONS

MANY GREAT GEOLOGISTS HAVE COME BEFORE US AND many will follow us. The great geologists will follow along the same hiking trail and, at some point in time, step off the path well-traveled and bravely pioneer a new path to a new understanding or develop into a memorable leader. Below are dedications of two geologists who deserve to be memorialized for their work and efforts. Diane Murbach presents the following dedications:

WERNER RAPHAEL LANDRY
1946–2017

Werner served as an officer for SDAG, completing his volunteer geologic service as our Past President in 1998. Werner organized our annual 1997 geologic field trip, and was an editor for our 1997 publication titled "*Santa Cruz Island Geology Field Trip Guide*," edited by James R. Boles and Werner Landry.

Werner was born in Germany and he moved to the US when he was 10 years old. He joined the Marines and served in Vietnam as a radar technician. He returned to San Diego and attended SDSU where he earned his degree in Geology. He enjoyed a long career as an engineering geologist spending the last 17 years working for the City of San Diego before retiring in 2009. His passion for the outdoors took him around the world. He was an accomplished rock climber and mountaineer.

JAMES RICHARD EVANS
1942–2016

Jim served as an officer for SDAG, completing his volunteer geologic service as our Past President in 1989. Jim organized our annual 1988 geologic field trip, and was the editor for our 1988 publication titled "*Landsides In Crystalline Basement Terrain.*"

Jim was born on a small ranch 40 miles from Fort Casper, WY, near where his father managed an oil field. The Evans family moved to California in 1955. He entered SDSU as a Philosophy major, but all that changed when he walked into Baylor Brookes' geology class to fulfill a science requirement. He loved the field trips and mapping expeditions to Baja with his SDSU "geology family." He graduated with a geology degree from SDSU in 1965, and began an engineering geology career path that lasted for over fifty years

SDAG 2017 CORPORATE SPONSORS

DIAMOND (+$1,000)

HARGIS+ASSOCIATES, INC.
HYDROGEOLOGY • ENGINEERING

www.hargis.com
Contact: Dr. David R. Hargis

H&P Mobile Geochemistry Inc.

www.handpmg.com
Contact: Louise Adams or Suzie Nawikas

RUBY (+$500)

Tetra Tech, *www.tetratech.com*
Geocon, Inc., *www.geoconinc.com, Contact: Rupert Adams*
Trevet, Inc., *trevetinc.com*
Atlas, *www.atlasexc.com, Contact: Barry Anderson*

EMERALD (+$100)

Dr. Pat Abbott, *SDSU Prof. of Geology, Emeritus, sci.sdsu.edu/geology/emeritus*
Marty and Sherry Bloom
Joe Corones
Greg Cranham, *Consulting Geologist*
Damon DeYoung, *Battelle*
Dr. Margaret R. Eggers, *CHG-Eggers Environmental, Inc., www.eggersenv.com*
Karen Evans, *care of belated Jim Evans, past SDAG president, circa 1989*
Phil Farquharson, *Geology Guy, geology-guy.com*
Carolyn Glockhoff, *Caro-Lion Enterprises, www.caro-lion.com*
Cari Gomes
Dr. Sarah Gray
Blayne Hartman, *www.hartmaneg.com*
Rob Hawk
Tom Lamb
Leighton Group, *Contact: Bob Stroh, www.leightongroup.com*
Lowell E. Lindsay and Diana Lindsay, *Sunbelt Publications/ AESE, sunbeltpublications.com*
Dr. Monte Marshall, *SDSU Prof. of Geology and Geophysics, Emeritus*

John Minch, *www.jma-ca.com*
James R. Ashby/Mission Geoscience, *Inc., www.missiongeoscience.com*
Jennifer Morton, *Terramar Environmental, www.terramar-environmental.com*
Monte and Diane Murbach, *Murbach Geotech, www.murbachgeotech.com*
Pacific Drilling Co., *www.pacdrill.com*
Peet's Coffee & Tea Solana Beach, *California*
Les Reed–Geotechnical Exploration Inc.
Rothman Creative *(Fritz Rothman)*
Gerald Shiller
Scott Snyder, *www.snydergeologic.com*
Southwest Geophysics Inc., *southwestgeophysics.com*
David and Jan Steller
Dr. Anne Sturz
Mary Walke
Carole L. Ziegler
Woody Higdon, *Geotech Imagery Intl., www.geo-tech-imagery.com*

The generosity of these sponsors allows our organization to keep membership dues low, reduce rates for students, produce monthly newsletters, purchase supplies, host the annual field trip, produce our geologic guidebooks, and provide scholarships to local geology students.

To contact any of the above sponsors, please visit www.sandiegogeologists.org.

2017 SAN DIEGO ASSOCIATION OF GEOLOGISTS
Rupert S. Adams: PRESIDENT
Chris Livesey: VICE PRESIDENT
Ken Haase: SECRETARY
Adam Avakian: TREASURER
Lowell Lindsay: PUBLICATIONS
Carolyn Glockoff: WEBMASTER

2017 SAN DIEGO GEOLOGICAL SOCIETY, INC.
PRESIDENT: Dave Bloom
VICE PRESIDENT: Bob Stroh
SECRETARY: Diane Murbach
TREASURER: Greg Cranham
DIRECTOR: Lowell Lindsay
DIRECTOR: Monte Murbach
DIRECTOR (*EX OFFICIO*): Rupert Adams

San Diego Association of Geologists is a program of San Diego Geological Society, Inc., a California 501(c)(3) nonprofit corporation.

SAN DIEGO ASSOCIATION OF GEOLOGISTS' PUBLICATIONS

1973 *Studies on the Geology and Geologic Hazards of the Greater San Diego Area, California*, Arnold Ross and Robert J. Dowlen, eds.

1974 *Recent Geological and Hydrologic Studies, Eastern San Diego County and Adjacent Areas: A Guidebook*, Michael W. Hart and Robert J. Dowlen, eds.

1975 *Studies on the Geology of Camp Pendleton, and Western San Diego County, California*, Arnold Ross and Robert J. Dowlen, eds.

1977 *Geology of Southwestern San Diego County and Northwestern Baja California*, Gregory T. Farrand, ed.

1978 *Natural History of the Coronado Islands, Baja California, Mexico*, Herman T. Kuper, ed.

1979 *Earthquakes and other Perils, San Diego Region*, Patrick L. Abbott and William J. Elliott, eds. (GSA field trip guide)

1981 *Geologic Investigations of the Coastal Plain, San Diego County, California*, Patrick L. Abbott and Shannon O'Dunn, eds.

1982 *Geologic Studies in San Diego*, Patrick L. Abbott, ed.

1985 *On the manner of Deposition of Eocene Strata in Northern San Diego County*, Patrick L. Abbott, ed.

1988 *Landslides in Crystalline Basement Terrain*, James R. Evans, ed.

1989 "Field Trip Road Log: Fault Features between La Jolla, U.S.A. and Ensenada, Mexico," by Diane Murbach: in *The Seismic Risk in the San Diego Region.*

1990 *Geotechnical Engineering Case Histories in San Diego County*, John H. Hoobs, ed.

1991 *Environmental Perils, San Diego Region*, Patrick L. Abbott and William J. Elliott, eds. (for a GSA annual meeting).

1992 *Natural History of the Coronado Islands, Baja California, Mexico* (revisited 1992), Lyne Perry, ed.

1993 *Colorado Desert and Salton Trough Geology*, Joe Corones, ed.

1994 *Geology and Natural History of Camp Pendleton, U.S.M.C. Base, San Diego County, California*, Phillip S. Rosenberg, ed. (2nd Edition, 2010).

1995 *Paleontology and Geology of the Western Salton Trough Detachment, Anza-Borrego Desert State Park, California*, Paul Remeika and Anne Sturz, eds.

1996 *Geology and Natural Resources of Coastal San Diego County, California*, Tissa Munasinghe and Phillip Rosenberg, eds.

1997 *Santa Cruz Island: Geology Field Trip Guide*, James R. Boles and Werner Landry, eds.

1998 *Geology and Geothermal Resources of the Imperial and Mexicali Valleys*, Lowell E. Lindsay and William Hample, eds.

1999 *Water for Southern California: Water Resources Development at the Close of the Century*, Gregory T. Cranham, ed.

2000 *Geology and Enology of the Temecula Valley, Riverside County, California*, Barbara B. Birnbaum and Kerry D. Cato, eds.

2001 *Coastal Processes and Engineering Geology of San Diego, California*, Robert C. Stroh, ed.

2003 *Geology of the Elsinore Fault Zone, San Diego Region*, Monte L. Murbach and Michael W. Hart, eds. (joint publication with SCGS)

2004 *Mining History and Geology of Joshua Tree National Park*, Margaret R. Eggers, ed.

2005/06 *Geology and History of Southeastern San Diego County, California*, David M. Bloom, Phillip T. Farquharson and Carole L. Ziegler, eds.

2006 *Mines and Geology of the Randsburg Area, An Historical Gem of the Mojave Desert*, by D.D. Trent.

2008/09 *Palms to Pines: Geological and Historical Excursions through the Palm Springs Region*, Jeff Miller, Bryan Miller-Hicks, and Scott Snyder, eds. (joint publication with SCGS).

2010 *Geology and Lore of the Northern Anza-Borrego Desert Region*, Monte L. Murbach and Charles C. Houser, eds. (joint publication with SCGS).

2011 *Picacho and the Cargo Muchachos*, Todd Wirths, ed.

2012 *Roadside Geology along Sunrise Highway*, by Michael J. Walawender

2012 *Waiting for Tsunami, Coastal Hazards of Northern San Diego County*, Cari Gomes, ed.

2013 *San Luis Rey on Display, Geoscience in Northern San Diego County, California*, Brian J. Olson, ed.

2014 *Coast to Cactus: Geology and Tectonics, San Diego to Salton Trough, California*, Jennifer Morton, ed.

2015 *Geology of the Coyote Mountains, Southern California*, Randall Wagner, ed.

2016 *Geology and Geologic Hazards of Northwestern Baja California: The Gold Coast and Agua Blanca Fault*, John Minch, Jim Ashby, and Rupert Adams, eds.

2017 *Julian-Banner Mining District: A Road to Gold*, Christopher D. Livesay, ed.

PART I

SUMMARY *of the* Field Trip Road Log

Field trip attendees (incomplete) at the entrance to the Eagle Mine, Julian, California. Photo courtesy of Monte Murbach.

FIELD TRIP Summary
DAY 1

START:

STOP 1 Ballena Gravels

Located adjacent and to the east of the San Diego County Fire Authority Station No. 85 (address: 25858 CA-78, Ramona, CA 92065, or more specifically at: 33.072854, –116.768344).

Park in the open space just south-southeast of the fire station and do not block the driveway from Highway 78 to the rear of the fire station.

Follow the power line access road to the top of the mountain.

Hike: 1.2 miles to top.

Hike: 1.2 miles to bottom.

STOP 2 Paleosol

Exit Stop 1 and turn left onto SR-78 and continue for 2.03 miles.

Turn right onto Old Julian Highway.

Continue 1.34 miles you will see a power pole on a roadside earthen platform.

Immediately beyond the power pole turn left into the Creek Hollow Ranch (east gate), Drive 100 feet and park off the pavement into the open field.

You will walk back to the power pole for further discussion.

As we stand on the side of the road, take care and attention to the passing traffic on Old Julian Highway. The weekend traffic is plentiful with cars and motorcycles traveling at high rates of speed.

STOP 3 West Preserve Hike:

At the power pole: backtrack 1.34 miles to SR 78.

Turn right onto SR 78 and continue east for 5.07 miles.

On your left will be the entrance to the County of San Diego Open Space West Preserve.

STOP 4 The Warrior Claim and Others

Continue east on SR 78 approximately 2.04 miles.

The turnoff is on your right and is designated as a picnic turnout.

STOP 5 Julian Mining Company

This Stop is skipped for the October 15–16 field trip.

STOP 6 Menghini Winery

Continue on SR 78 approximately 1.9 miles turn left on Wyonola Road.

Continue approximately 3.7 miles turn left on Julian Orchards Drive/Farmer Road; Menghini Winery will be up the road.

STOP 7 Elsinore Fault

This Stop is skipped for the October 15–16 field trip.

FIELD TRIP Summary
DAY 2

STOP 8 Ranchito Mine:

Leave Menghini Winery on Julian Orchards Drive and turn left onto Wynola Road.

Continue for 5.70 miles until you reach the convenience store located at 36342 CA-78 (Banner Grade Road), Julian, California.

Park at the Banner Grade convenience store and carpool in high clearance vehicles to Right Fender Ranch, which facilitates Ranchito Mine.

Exit Banner Grade Convenience store, turn right onto Banner Grade Road.

Back track on Banner Grade 100 feet and turn left onto a dirt road (California Riding and Hiking Trail).

Continue on the dirt road for 2.0 miles taking care to keep left.

You will come to one fork, keep left. At the second fork, the entrance to Right Fender Ranch will be on the left.

STOP 9 Chariot Canyon Mine

This Stop is skipped for the October 15–16 field trip.

STOP 10 Eagle Mine: 9.75 mi

Turn around and descend back to Banner Grade, keep right until you reach Banner Grade Road: 2.0 miles.

Turn right onto the paved road (Banner Grade) and return to your vehicle, don't forget to stop and have some refreshments at the Banner Grade Convenience store.

From the convenience store, turn right onto Banner Grade Road and drive up gradient towards downtown Julian.

Just as you enter downtown Julian, turn right onto C Street and follow signs to Eagle Mine.

Have lunch and tour.

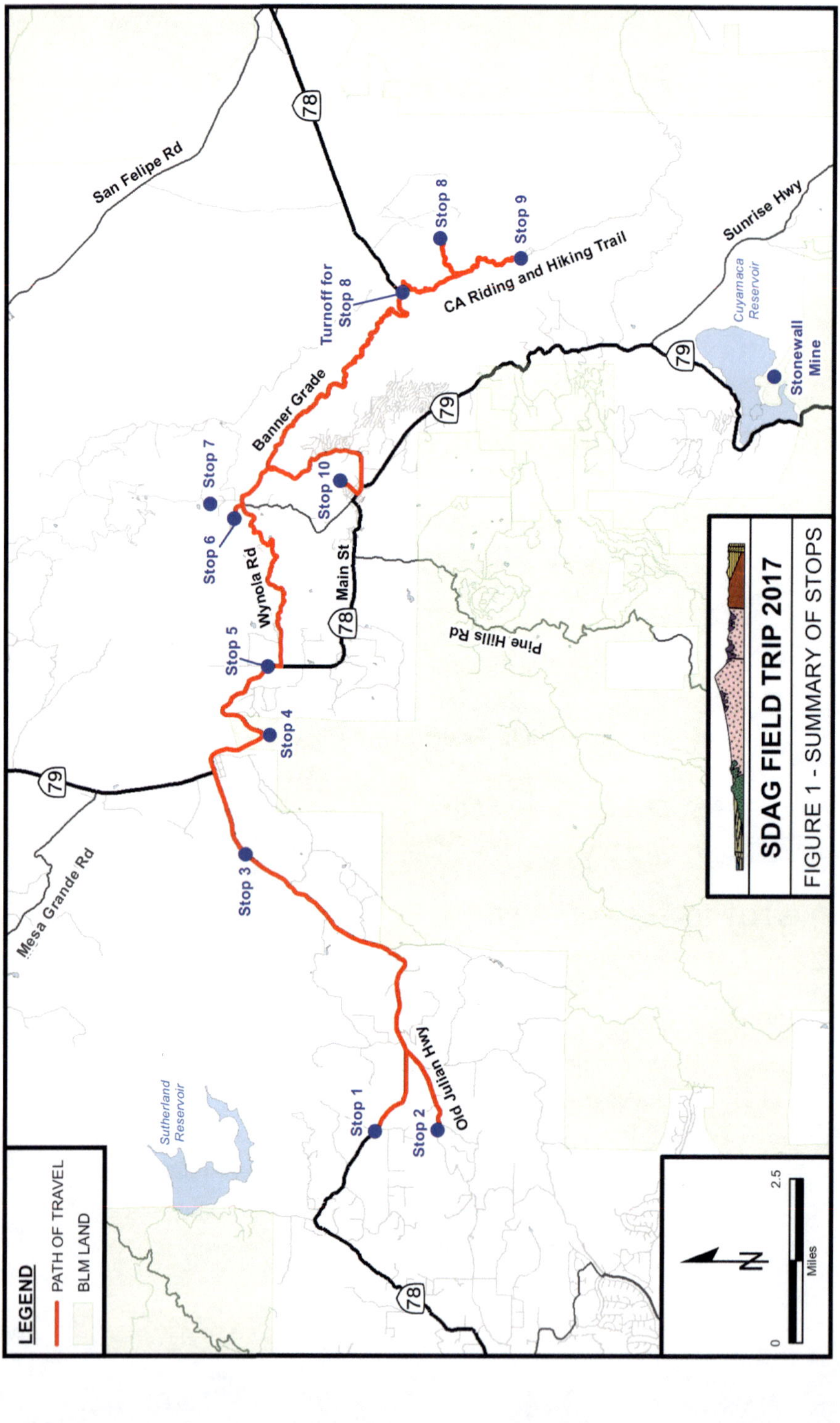
SDAG FIELD TRIP 2017
FIGURE 1 - SUMMARY OF STOPS
LEGEND
PATH OF TRAVEL
BLM LAND
0
2.5
Miles
N
San Felipe Rd
78
Stop 8
Stop 9
Sunrise Hwy
CA Riding and Hiking Trail
Turnoff for Stop 8
Cuyamaca Reservoir
79
Stonewall Mine
Banner Grade
Stop 7
Stop 10
Stop 6
Wynola Rd
Main St
Pine Hills Rd
Stop 5
Stop 4
Stop 3
Mesa Grande Rd
Old Julian Hwy
Stop 1
Stop 2
Sutherland Reservoir

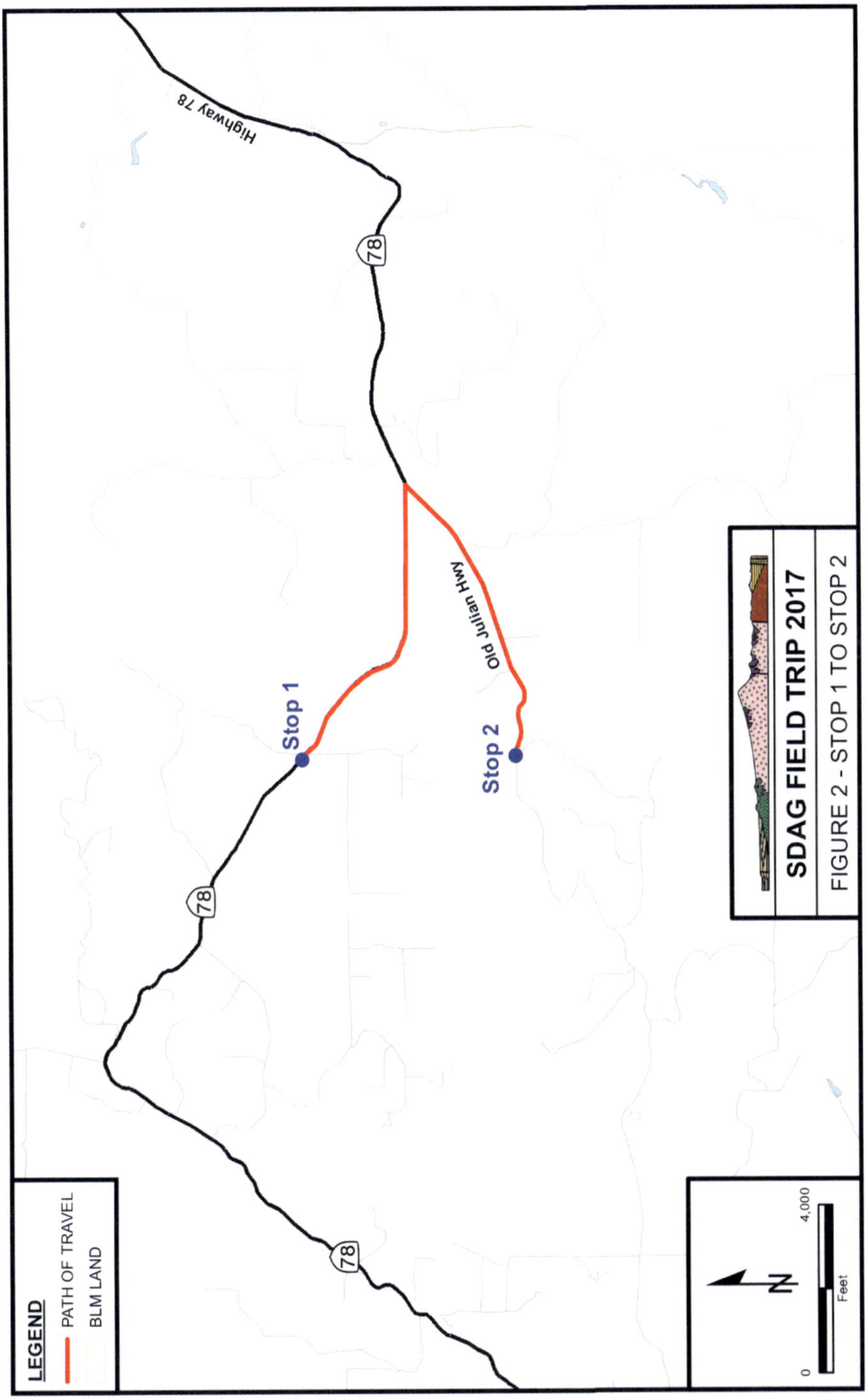
Highway 78
78
Old Julian Hwy
Stop 1
Stop 2
78
78
LEGEND
PATH OF TRAVEL
BLM LAND
SDAG FIELD TRIP 2017
FIGURE 2 - STOP 1 TO STOP 2
N
0
4,000
Feet

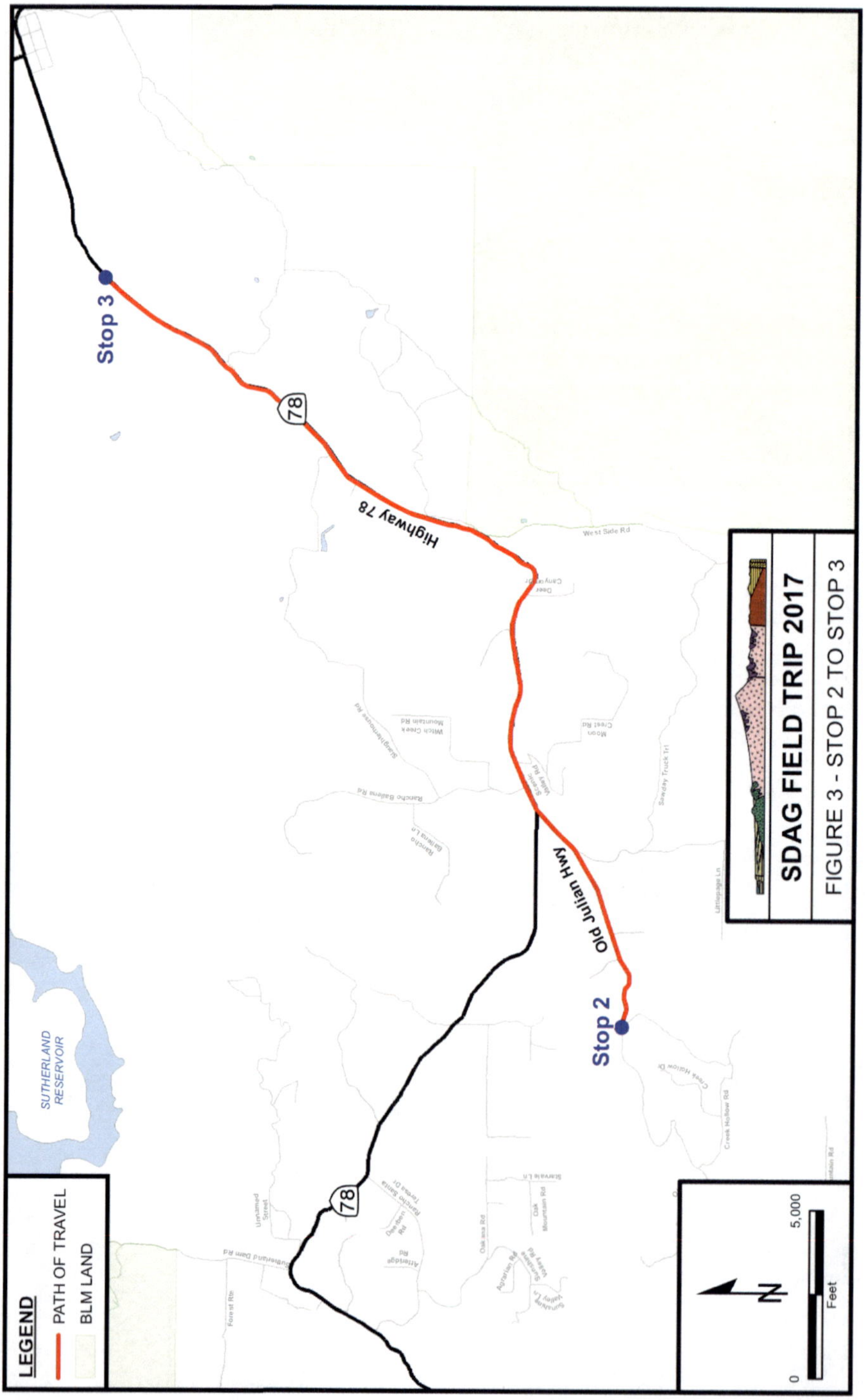

FIGURE 3 - STOP 2 TO STOP 3

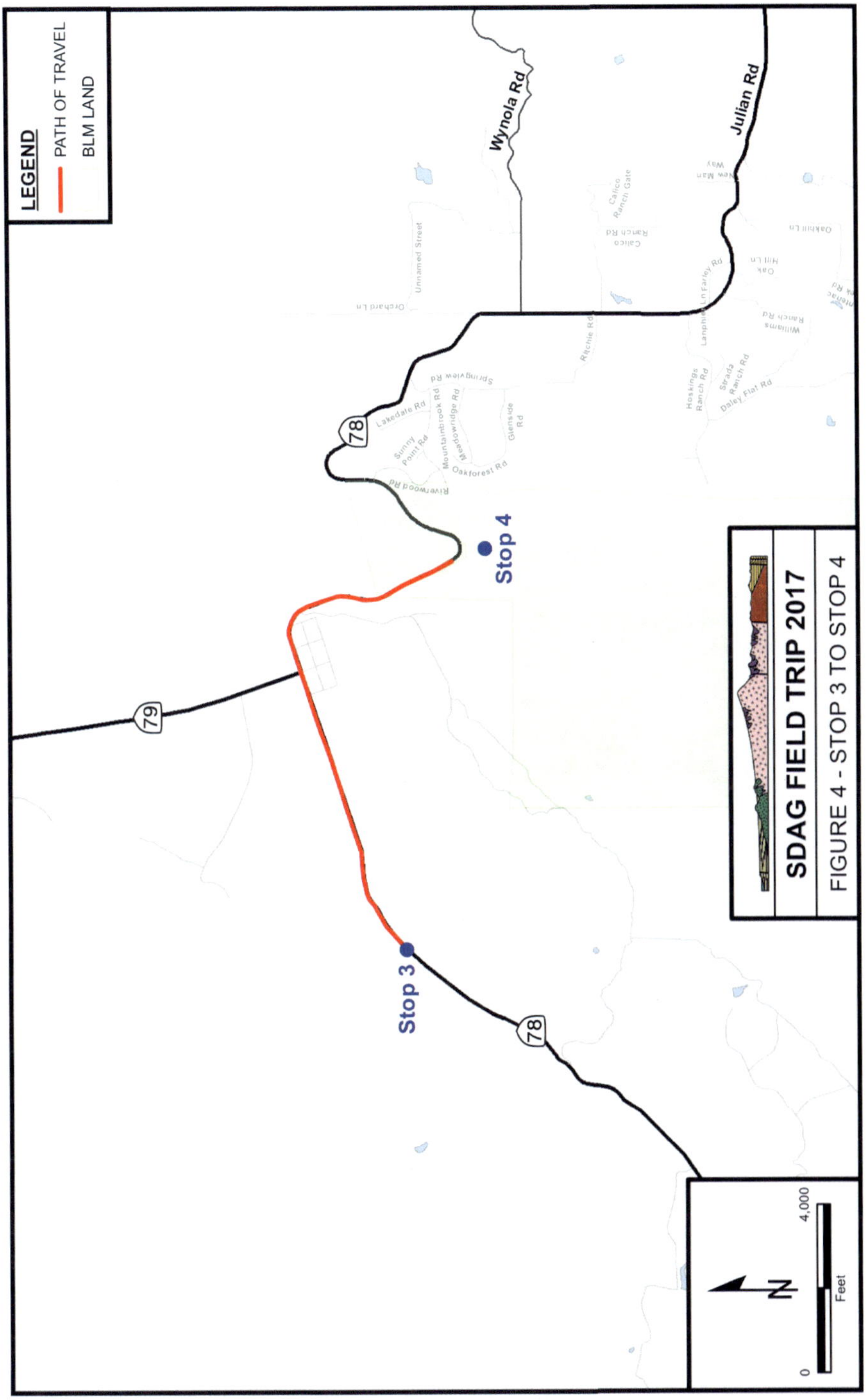
LEGEND
PATH OF TRAVEL
BLM LAND
Wynola Rd
Julian Rd
78
79
Stop 3
Stop 4
SDAG FIELD TRIP 2017
FIGURE 4 - STOP 3 TO STOP 4
N
0
4,000
Feet

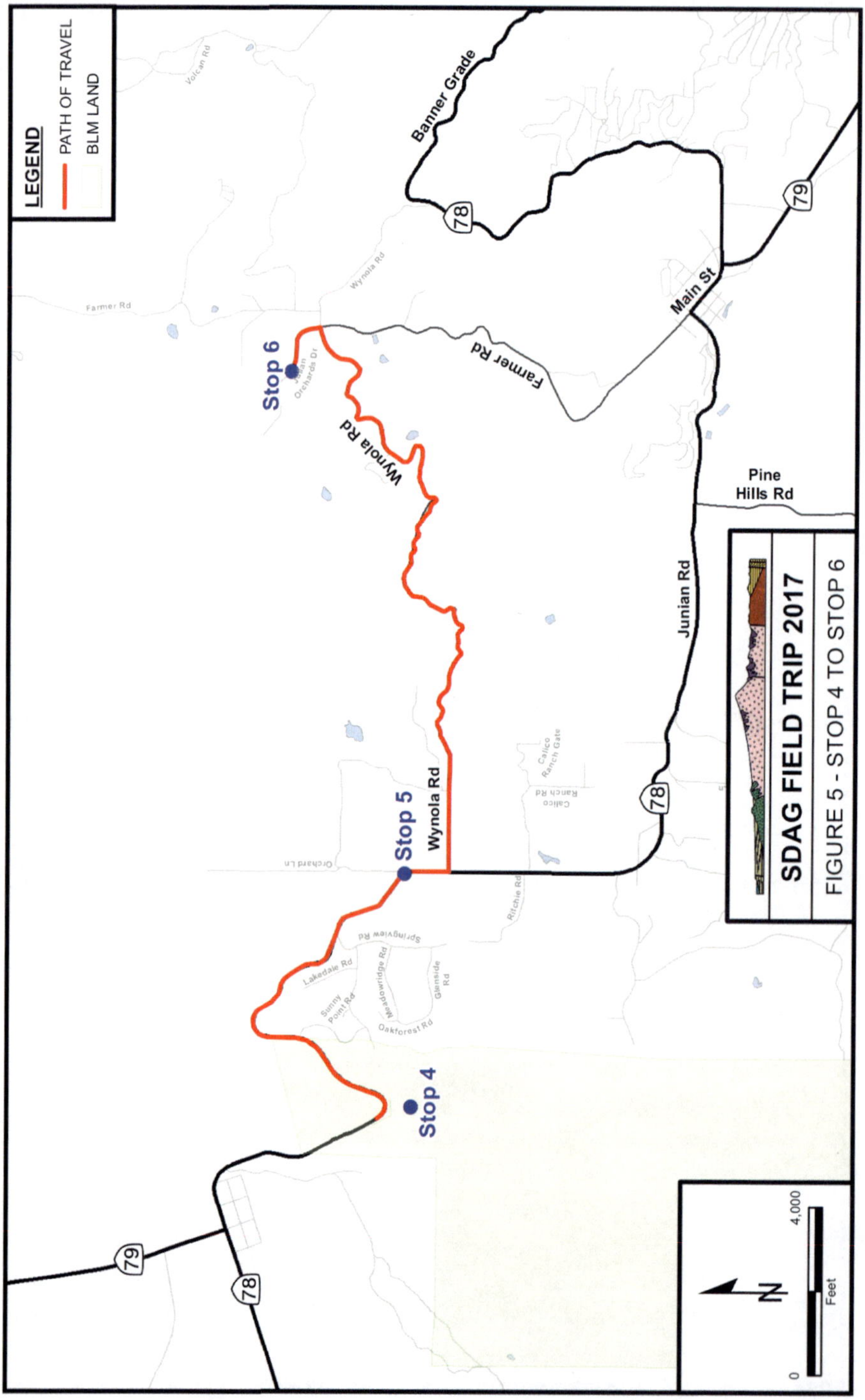

FIGURE 5 - STOP 4 TO STOP 6

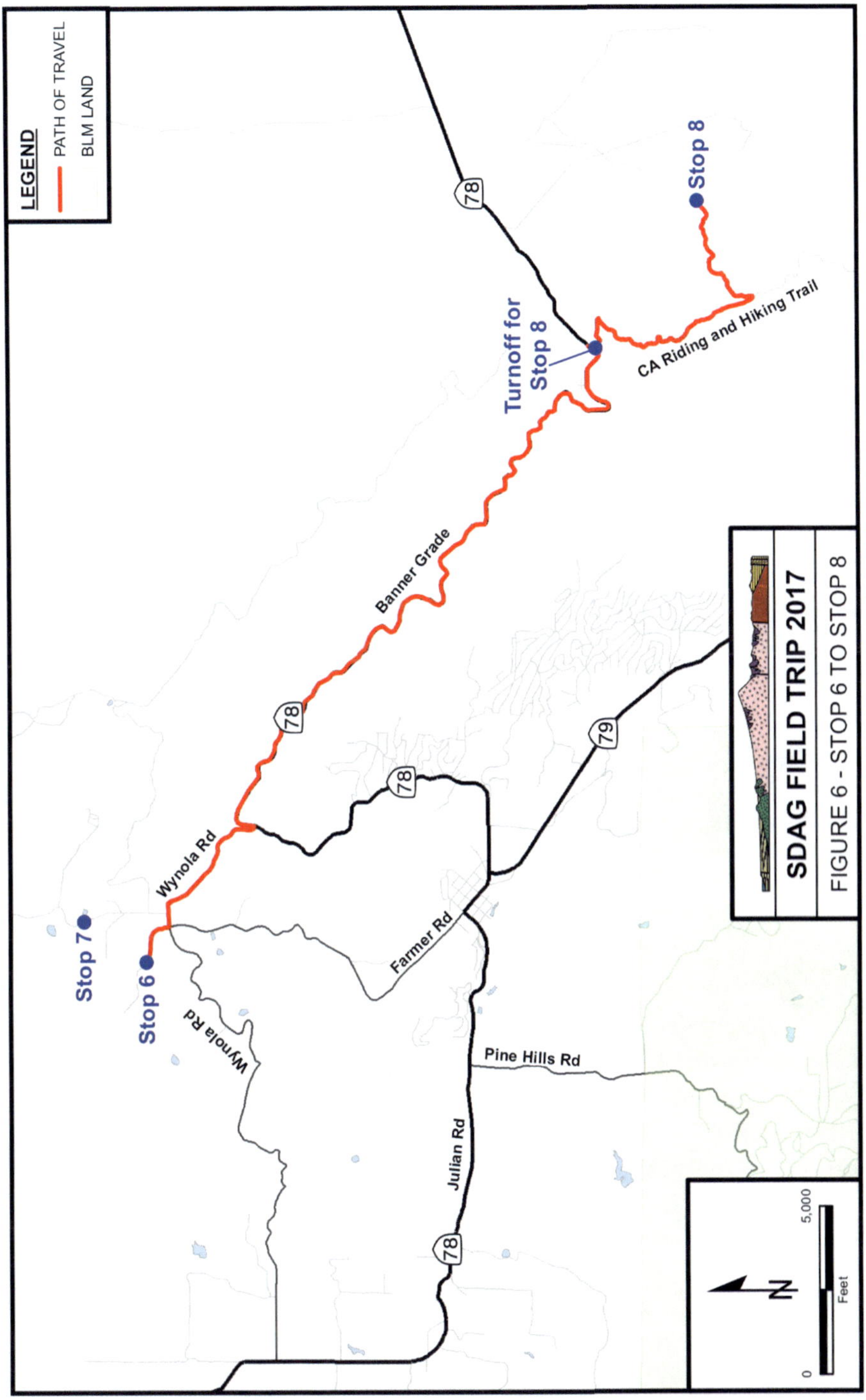

FIGURE 6 - STOP 6 TO STOP 8

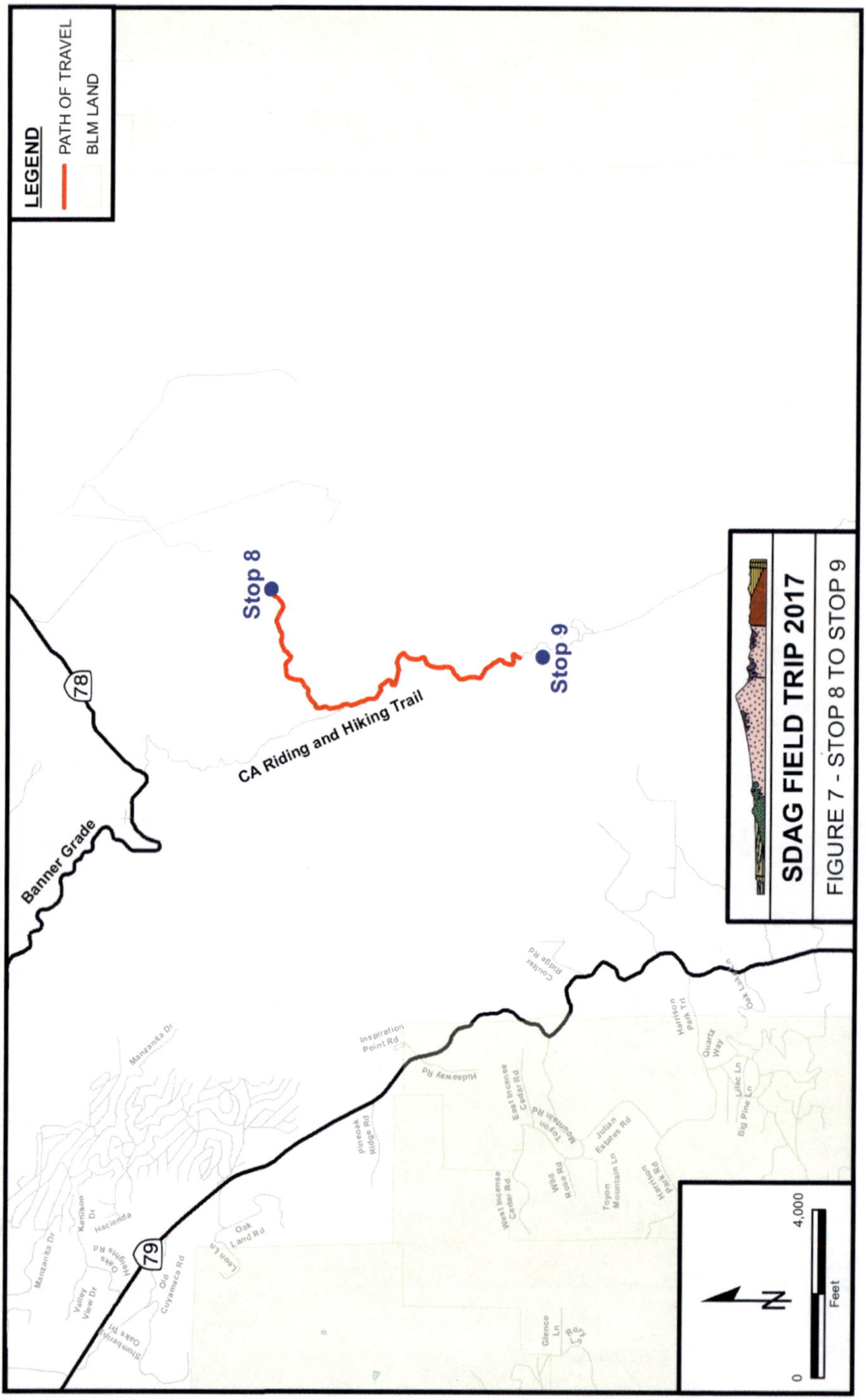

FIGURE 7 - STOP 8 TO STOP 9

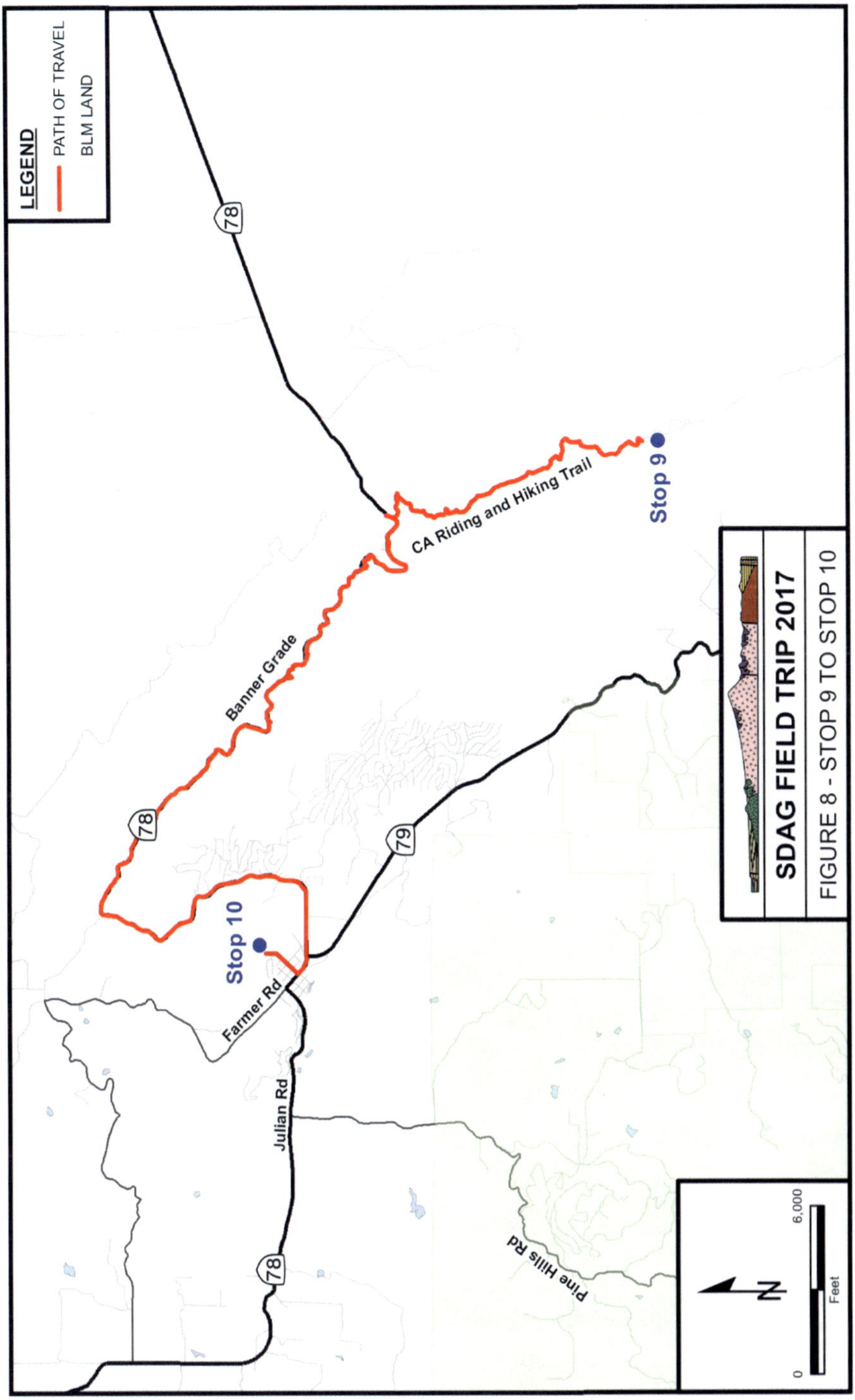

SDAG FIELD TRIP 2017

FIGURE 8 - STOP 9 TO STOP 10

PART II

DETAILED
Field Trip Road Log

FIELD TRIP DAY 1

START/STOP 1: Ballena Gravels

Hike Leader: Dr. Patrick Abbott

Hike: 1.2 miles to top, 1.2 miles to bottom.

Stepping outside of your vehicle, take a deep breath of fresh crisp air (the crisp air is most enjoyable in the morning or shortly after a rain event, as a chicken/egg farm is nearby and Stop 1 is located downwind of the chicken/egg farm).

You have arrived from various places, wherever you may come from; from the stir-crazy downtown of San Diego, or one of the many suburbanized outskirts, today you will embark into the fresh mountain air.

Stop No. 1 (Ballena Gravels) is located adjacent and east of the San Diego County Fire Authority Station No. 85 (25858 CA-78, Ramona, CA 92065, or more specifically at 33.072854, -116.768344).

This is a hike-in/hike-out, 1.2 miles each way with a total round trip of approximately 2.4 miles with a net elevation gain of approximately 500 feet.

Just northeast of the fire station, a dirt road facilitates an easy hiking path to the destined endpoint. As you hike up hill, you will notice massive Cretaceous-aged tonalite underlying a surface littered with low-grade minerals (quartz, feldspars, micas, etc.); however, persevere and continue. In approximately ½ mile the road will fork to the right. Proceed to the right and continue. After an approximate ½ mile, just before the last ascent to the top, you will notice a geologic transition from Cretaceous-aged tonalite to an Eocene-aged conglomerate. The mystery of the gravels precariously atop a 3,000 foot hill top is best explained by Professor Patrick Abbott. The site-specific paper can be found in Part III of this guidebook.

Take a few moments to take in the view, read Pat Abbott's paper, and appreciate the magnitude of geologic scale and time.

Then return to your vehicle.

STOP 2 Paleosol

Exit Stop 1 and turn left onto SR-78 and continue for 2.03 miles

2.03 miles Turn right onto Old Julian Highway and continue for 1.34 miles.

3.37 mile You will see a power pole to your left along the roadside, park along the shoulder of the highway, but be aware that Old Julian Highway is a popular road for sports cars, motorcyclists and bicyclists. Take caution, as vehicles might be traveling at a high rate of speed.

In 2012, this site, among others, was suggested to present evidence of an ancient tsunami deposit generated by a meteorite impact with earth. Features at this site will be discussed that do not support evidence for a tsunami deposit. To understand the full content of the new interpretation represented at Stop No. 2, take a few minutes to read the paper written by Bill Elliot. The following is an excerpt from Bill Elliot's article specific to Stop No. 2:

> *The power pole site* (Site 3) is exposed in road cuts on both sides of Old Julian Highway (*Figure 1, Table 1, Elliot, 2017*), readily identified by a power pole on the south side of the road (*Photo 3, Elliot, 2017*). Here we note the presence of buried paleosols (though less well developed) and *in situ* mud cracks similar to the Site 1 and 2 arroyo exposures. Grussified granitic clasts are present—attesting to periods of intense weathering. Although "negative evidence," we nowhere found stratigraphic features even vaguely suggestive of a paleo-tsunami, much less the ~2.51 Ma Eltanin event. (Elliot, 2017)

STOP 3 West Preserve Hike:

From the power pole you will back-track on Old Julian Highway onto SR 78

1.34 miles Turn right onto SR 78, continue for 5.07 miles.

6.41 miles On the left will be the entrance to the County of San Diego Open Space West Preserve.

Mike Thacker has provided a colorful discussion of the geology on a local and regional scale, with interdisciplinary comments. In the center of the parking lot, there is picnic table. Sit at the table and take a few minutes to read "A Geological Hike Into The Santa Ysabel West Open Space Preserve" presented in Part III of this guidebook.

After reading the article, follow in the footsteps to more fully understand the local geology. Return to your vehicles.

STOP 4 The Warrior Claim and Others

From Stop 3 continue eastbound on SR 78.

2.04 miles On your right, there will be a turnoff for a designated picnic rest stop.

The picnic area encroaches into Federal territory managed by the Bureau of Land Management (BLM). In the vicinity of the picnic area, there are 3 placer mining claims. Gold Prospectors No. 1 and No. 2, as well as, Warrior. Although the claims are on public land, the claims provide mineral rights to the claim owners. That being said, you will need to contract or gain permission from the claim owners prior to prospecting within the claim boundaries. However, the Gold Prospectors claim is owned by the Gold Prospectors Association of America (GPAA) and if you hold current membership, the GPAA allows you to prospect on their claims (this may change with time, perform your due diligence). To better help you understand active claims the BLM has a publicly available data base known as the BLM LR2000:

> The Bureau of Land Management (BLM) administers more public land—over 245 million surface acres—than any other Federal agency in the United States. Most of this land is located in the 12 Western states, including Alaska. The BLM's Legacy Rehost System, called LR2000, provides reports on BLM land and mineral use authorizations for oil, gas, and geothermal leasing, rights-of-way, coal and other mineral development, land and mineral title, mining claims, withdrawals, classifications, and more on federal lands or on federal mineral estate. The LR2000 enables users to access the same information over the Internet as in the BLM State Office Information Access Centers. [*The LR2000 can be found on the BLM's webpage or the following address: https://reports.blm.gov/reports.cfm?application=LR2000 (Bureau of Land Management, 2017)*]

Inaja Picnic Rest Stop:

The following excerpt is from the official website for the County of San Diego Parks and Recreation Department on the Inaja Memorial Picnic:

> The Inaja Memorial Picnic area was established to memorialize the 11 firefighters who lost their lives battling the 60,000-acre Inaja Fire in 1956. With an elevation of 3440, the higher altitude offers year round cooler temperatures, and great views of surrounding

> mountain ranges. Besides the beautiful mountain scenery, this site offers a quiet tranquil setting that creates a great environment for afternoon family picnics, with connecting trails, interactive site seeing activities and scenic areas. For more information on the Inaja and connecting trails, please see our *Inaja Trail Recreation* page [https://www.fs.usda.gov/recarea/cleveland/recarea/?recid=47548]. For a more detailed look at this trail, read our Spotlight: Inaja Trail [https://www.fs.usda.gov/detail/cleveland/news-events/?cid=FSEPRD481881]. [https://www.fs.usda.gov/recarea/cleveland/recarea/?recid=47546, US Department of Agriculture, 2017.]

To view panoramic sceneries, take the trail beyond the restroom facilities, which will abruptly fork to the right, keep right and wander uphill. You will wrap around the hill above the parking lot. Take time to inhale frequent breaths of local aromatic shrubbery. At the top, you will find an eye-scope attached to a swivel azimuth. The azimuth will point you in the direction of places that are known regionally, but through zero-magnification, are far off.

After you are done embracing the breathtaking views, wander back to the parking lot and take time to look eastward across SR 78. As you might notice, the hillside evidently has been cut down approximately 20 feet to allow for the construction for the road alignment through the ridge that most surely was present prior to the road's existence. As a result, along the inside lane a cross cut of the subsurface geology is exposed at the surface.

Geologists admire these road cuts to the most; and as we shall discuss, this road cut tells a story, a story open to your own interpretation. For a starting proposition I will provide you with my version of the story.

For proper orientation, SR 78 wraps around the natural ridge with the tangent of the apex trending in a north-south direction, so as you stand looking at the road cut, to your left is north and to your right is south.

Keep standing at a distance to review the road cut from north to south, then south to north, then repeat until clarity formalizes. Clearly the geology is different in the north when compared to the south.

Preliminary geologic mapping by Janis L. Hernandez and Victoria R. Todd suggest the exposed units are similar to facies mapped and described in the El Cajon 30′ × 60′ geologic map (Todd, 2004).

The southern, unit-Kgm, exposes massive, coarse grained (~1 cm) felsic mineral grains, with lesser iron and muscovite constituents, the unit is very weathered and expresses some foliage and jointing.

Panning to the north, unit-Kpv, is composed of coarse grained (~½ cm) quartz and mafic minerals (hornblende), fresh exposure (non-weathered), with veins of quartz.

Within Unit-Kpv are abundant mafic xenolyths (or inclusions depending on your definition) and a possible large schlearan, nearly truncated by the non-conformity separating unit-Kpv from Unit-Kgm. The contact bisects the road cut and dips to the south. Panning farther east (above the road cut and along the natural topography of the ridge). There appears to be more resistant weathering on the northern side of the projected contact with residual boulders and raised topography in comparison with the groundsurface to the south of the contact, indicating Unit-Kgm is more susceptible to uniform weathering and erosion.

Further east is a prominent outcrop of Julian Schist indicating as we traverse east and away from the Ramona Plutonic Complex, mixing of Julian Schist with igneous intrusions becomes more large scale in comparison to the mixing we viewed at the edge of the RPC of the Santa Ysabel Open Space west Preserve.

Photo by Christopher D. Livesey

STOP 5 **Julian Mining Co.,** 4444 CA-78, Julian, CA 92036
Continue eastbound on SR 78 approximately 1.89 miles.

1.89 miles There will be a sign on your left, just after a bend to the right. Here you can learn more about the local mining and practice your gold panning for a nominal cost.

STOP 6 **Menghini Winery,** 1150 Julian Orchards Drive, Julian, CA 92036
Continue on SR 78 approx. 300 feet, turn left on Wyonola Road.

300 Feet Turn Left on Wynola Road, continue approximately 3.7 miles

3.7 miles Turn left on Julian Orchards Drive / Farmer Road.
Menghini Winery is the dedicated camping spot for the field trip. If you are reading this after the trip has occurred, please enjoy the winery. The owners are very nice and enjoyable. Enjoying a glass of wine on their back porch overlooking their sprawling views is satisfying after a long day of activities. Leaving Menghini Winery, there are a few other public camping spots available, such as William Heise County Park, Pinezanita RV and Campground, Lake Cuyamaca and Mount Laguna.

Photo: Oblique Google Earth photo, overlain with GIS fault map data showing the Elsinore Fault (in red) from California Geologic Survey. (https://maps.conservation.ca.gov/cgs/#datalist and Google Earth, 2017)

STOP 7 Elsinore Fault

This stop is not intended to further understand the mining of the trip, however the feature might be notable to some. A trace of the Elsinore fault transects Farmer Road in a road cut at the approximate location of: Latitude North 33°06′32″, Longitude West 116°36′10″.

An eager geologist will need to spend time scraping the roadcut for a fresh surface, as the surface is obscured by seasonal weathering.

FIELD TRIP DAY 2

STOP 8 Ranchito Mine

Leave Menghini Winery on Julian Orchards Drive and turn left on Wynola Road. Continue for 5.70 miles. This route will bypass the town of Julian, California. However, do not be discouraged, this road log will route you into Julian after surveying (from a distance) the Ranchito and Golden Chariot mines. Below, Lowell Lindsay eloquently summarizes Banner Grade and associated interests:

GOLDEN DREAMS ON GRANITE MOUNTAIN
Banner Grade, Ranchito Mine, Rodriguez Canyon
Lowell Lindsay, Sunbelt Publications, 2017

Beginning in the piney mountain village of Julian (elev. 4220′), Hwy 78 jackknifes steeply down seven miles of Banner Grade to the 1870s gold rush town site of Banner (elev. 2,760′) in the high desert at the foot of massive Granite Mountain (elev. 5633′). Geology enthusiasts (not including drivers) will see clear visual evidence of the great Elsinore Fault on canyon slopes opposite (northeast of) the highway. Trace exposure of the Elsinore Fault can also be observed from a distance (later) as you ascend into Chariot Canyon, or near the Right Fender Ranch (Stop No. 8).

A final hairpin turn near the bottom of the grade approaching Banner Store marks Chariot Canyon where a major trace of the Elsinore Fault crosses the highway and continues 50 miles southeast to the Coyote Mountains near the Mexican border. This canyon was a focus of the Julian Gold Rush of the 1870s where mines and prospects pockmark the steep hillsides and stream bottoms. Almost all of the area is restricted as private land, so the mine adits of golden dreams must be enjoyed from roadside views.

A rocky dirt road, just uphill (west of) Banner Store leads steeply up southwest into the Chariot and Rodriguez Canyon backcountry. (It may be "dummy locked" but not latched for animal control.) As with any backcountry gate that is opened, the driver is responsible to close it securely behind themselves. The road forks after a 1.25 mile climb with the right branch leading south and down into Chariot Canyon and up into Anza-Borrego Desert State Park.

The left fork turns east and continues the ascent over a ridge and turns southeast toward the saddle of Rodriguez Canyon. It is along here that that the "just right" Goldilocks mix of gold in Julian Schist, matrix of Quartz Diorite, and dynamic faulting produced lucrative deposits with lilting names like Elevado, Aguajito, Ranchito, and Desert Queen.

Huge live oak trees and pastoral scenes soon reward the traveler with views of the Ranchito Mine and the Right Fender Ranch, now owned since 1974 by Chris E. Richardson whose family heritage in this area extends deep into the colorful history of the American Southwest and a famed frontier colonel named Couts.

The Ranchito Mine and its large, and classic five-stamp mill, rate high among the historical treasures of this area. Ownership of this mine came into the family of Colonel Cave J. Couts, a US Army officer who figured prominently in San Diego history during and immediately after the Mexican War. His diary records the first of the rush to the California gold fields after the close of war in 1849. Couts' son, Cave J. Couts Jr., found his own golden bounty when he purchased the Ranchito Mine in 1895 for $5,500 cash. The Ranchito produced about 2,500 ounces of pure gold and about 100 ounces of silver between 1895 and 1948. Considering the price of gold today, that would approach a million dollars. One might surmise correctly that the Richardson lineage was and is fortunate indeed.

Protection and respect for these natural and historic lands earns the fortune for us all.

REFERENCES AND SOURCES

CDMG County Report #3, San Diego, 1963

Anza-Borrego A-Z, D.E. Lindsay, 2000

Anza-Borrego Desert Region, Lindsay & Lindsay, 1978–2017

A Good Camp: Gold Mines of Julian and the Cuyamacas, Leland Fetzer

Personal interviews: C.E. Richardson, Monte Marshall

Notes from LEL & T. Wirths on SDAG annual field trip, 2017-10-15

5.70 miles Banner Recreational Ranch Convenience Store located at 36342 CA-78 (Banner Grade Road), Julian, California.

Park at the Banner Grade Convenience Store and carpool in high-clearance vehicles to Right Fender Ranch that facilitates Ranchito Mine.

From Banner Grade Convenience Store to the Ranchito Mine, turn right onto Banner Grade Road for 100 feet.

100 feet Turn left onto a dirt road (California Riding and Hiking Trail).

1.25 miles Turn left.

2.00 miles You will approach Right Fender Ranch.

You will need permission to enter Right Fender Ranch. Permission is unlikely and you should not enter. However, mine tailings can be seen in the distance to the northeast.

The following excerpts from the San Diego Association of Geologists 2010 Guide Book *Geology and Lore of the Northern Anza-Borrego Desert Region*:

> The Ranchito ore deposit crops out low on a southwestern facing ridge slope, which partially opens into a small valley. The deposits are enclosed in hybrid ("mixed") rocks, which are composed of quartz diorite and schist. As the deposit lies on the northeast side of the Elsinore Fault, which extends along the base of the ridge slope, the hybrid rocks are highly sheared. The ore deposit consists of quartz bodies in shear zones, which strike west-northwest along the fault (Weber, 1963). (Murbach, 2010)

STOP 9 Chariot Canyon Mine

Leave Right Fender Ranch and return to the California Riding and Hiking Trail.

0.75 Miles Merge onto the California Riding and Hiking Trail and continue ascending up the canyon.

As you ascend up the canyon, several side roads fork to the right that lead to adits of individual prospecting claims. Do not take any of these side roads.

1.92 miles As the road bends to the left, a side road to the right will take you down to the Chariot Canyon Mine. A gate inhibits trespassers and access will need to be granted prior to entering. In 2010, the San Diego Association of Geologists organized a day trip to the mine under previous ownership. However, the land changed ownership in 2011, and access could not be granted. Nevertheless, information regarding the mine and geology are presented herein this guidebook under Part III ("Selected Geological and Historical Aspects of the Julian Gold Mining District," page 139).

After Redman and the Baileys made their claims in the canyon, George V. King began his search in February 1871. He went further up the canyon to discover one of the richest veins in the Banner area and named it the Golden Chariot Mine. The canyon and later the mining district took its name from this min. It was the richest strike in southern California and produced about $2 million dollars worth of bullion.

The Chariot Canyon fault at this location formed along the eastern boundary of the transition zone between the eastern and western zones of the Peninsular Ranges Batholith. The transition zone is made of strongly deformed gneissic rocks and Late Triassic-early Jurassic Julian Schist. Together these rocks make the plastically deformed Cuyamaca–Laguna Mountains Shear zone (CLMSZ). Rocks to the east of the CLSMZ, according to recent papers by Grove, et al, (2003), and Bethyl-Thompson, et al, (in preparation) were originally thrust westward over western and transition zone rocks along moderately east-dipping faults; the Sunrise Highway fault and the Chariot Canyon fault. There is some evidence that this westward movement was reversed during the Miocene. As a result, on the Chariot Canyon fault there was significant down-to-the-east movement that accounts for the prominent escarpment along the east side of the range. (Murbach, 2010)

STOP 10 Eagle Mine

2320 C Street, Julian, CA 92036

Turn around and go back down grade to the intersection with Banner Grade, stay on main dirt road as minor side roads venture off to prospecting claims.

2.0 miles You will reach Banner Grade / SR 78

Turn right on the paved road (Banner Grade) and return to vehicles, don't forget to stop and have some refreshments at the Banner Grade Convenience Store.

From the convenience store, turn right and drive up gradient on Banner Grade.

9.75 miles Turn right on C Street

Continue on C Street several hundred feet and follow signs to Eagle Mine.

Have lunch, practice gold panning, and tour.

SUPPLEMENTAL ACTIVITIES IN THE JULIAN AREA

Hopefully you have enjoyed traversing the route taken during the 2017 San Diego Association of Geologists fieldtrip. There are other noteworthy places of interest in Julian, California and the surrounding area. Below is a summarized list of activities. Please note, that research should be performed prior to planning to assess the following businesses and or locations are active, have availability by reservation or by non-appointment, days and hours of operations, or require special conditions, etc.

SEASONAL APPLE PICKING

Volcan Valley Apple Farm
1284 Julian Orchards Drive, Julian, California
760–302-4574

Apple Starr Orchard
1020 Julian Orchard Drive, Julian, California
760–305-2169

Apple Lane Orchard Estate
2641 Apple Lane, Julian, California
760–765-2645

MUSEUMS

Julian Pioneer Museum (limited hours)
2811 Washington Street, Julian, California
760–765-0227

Julian Historical society museum (above Pioneer Museum, limited hours)
2133 4th Street, Julian, California
760–765-0683

OUTDOOR ACTIVITIES

Horseback Riding (ramona/Hollow Creek Ranch)
25279 Creek Hollow Drive, Ramona, California
760–789-4875

Volcan Mountain trailhead
1107 Farmer Road, Julian, California
760–765-4098

SHOPPING IN DOWNTOWN JULIAN, CALIFORNIA AND BEYOND

Several stores front Main Street where the traveler can sit to enjoy lunch, a refreshing beverage, and shop for antiques or unique items. Quaint store fronts continue approximately 6 miles west of Julian in the town of Wynola, California. Approximately 19 miles farther west resides the city of Ramona, where the traveler can find a mixture of specialty shops and modern day amenities.

REFERENCES:

Todd, Victoria R.; Preliminary Geologic Map of the El Cajon 30′ × 60′ Quadrangle, Southern California, Version 1.0, digital preparation by Rachel M. Alverez and TGS (Techni Graphic System, Inc.)

Murbach, Monte L. and Charles E. Houser; 2010; Geology and Lore of the Northern Anza-Borrego Desert Region: The Lows to Highs of the Anza-Borrego Desert State park; Sunbelt Publications.

Elliot, William and Roy J. Shlemon; 2017, Origin of "Pseudo-Tsunami" Deposits in the Ramona Area, San Diego County, California.

https://reports.blm.gov/reports.cfm?application=LR2000 , July 20, 2017.

https://www.fs.usda.gov/recarea/cleveland/recarea/?recid=47546, August 4, 2017

https://maps.conservation.ca.gov/cgs/#datalist , October 29, 2017.

PART III

SUPPLEMENTAL Papers

Ballena Hill

Remnant of a Golden River from the Eocene

PATRICK L. ABBOTT

Author of The Rise and Fall of San Diego

ON 26 APRIL 2008, AN SDAG ONE-STOP WONDER TRIP to Ballena Hill was organized by Mike Hart. Discussion of the Eocene fluvial system and its Poway rhyolite clasts made for an enjoyable and worthwhile Saturday morning. A highlight of the adventure was provided by George Copenhaver; he brought his pans and a supply of water then separated a significant amount of gold from the Ballena Gravels while we were on site.

SIERRA NEVADA GOLD

The placer gold fields of the Sierra Nevada foothills are famous worldwide. They formed when gold eroded from eastern sources was carried westward in Paleogene rivers and then deposited, especially in Eocene river-channel sediments in what is now the Sierra foothills. In the late 1800s, these Eocene auriferous fluvial conglomerates and sandstones were blasted apart via hydraulic mining. Forceful jets of water aimed at outcrops disaggregated the sedimentary rocks thereby releasing their particles of gold.

It is now recognized that an extensive Eocene paleoriver system drained much of what is now the western Great Basin into the Pacific Ocean. Water flowed westward from an elevated Nevadoplano across a lower elevation Sierra Nevada and on into the Pacific Ocean (Henry, 2009).

Eocene river deposits in the Sierra Nevada foothills contain a lot of gold. Do the foothills of San Diego have similar auriferous conglomerates of Eocene age? Yes, they are the Ballena Gravels.

SAN DIEGO COUNTY

The Ballena Gravels were recognized and well described by H.W. Fairbanks in 1893 in his report on "The Geology of San Diego County". Giving credit where credit is due requires a rather lengthy quotation from Fairbanks; the quotation below has been shortened.

> It is not generally known that an ancient auriferous gravel channel exists in the county. It begins about … 3 miles west of Ballena post office, where there rises a hill shaped like a whale's back (hence the name Ballena), covered with washed gravel and boulders. The main portion of the channel which has escaped erosion begins … capping a hill which has an elevation of 2,400 feet … The gravel is 50 to 100 feet thick, and has a width of 2,000 feet or more. It rises 300 to 500 feet above the valleys and canyons on its sides. It extends in a direction a little south of west for about 4 miles … A pretty valley, a mile long, has been eroded in the eastern end of the gravels, down to the underlying granite. Placer mining has been carried on for years here in a small way by Mexicans. Gold is said to be scattered everywhere through the gravels, which are often very firmly cemented. Lack of water, for the ridge is higher than any of the surrounding country, has prevented work on a large scale. Lately a mining district has been organized, and it is proposed to bring water 7 miles in a pipe. In the gravels are washed boulders, many of them being 2 feet in diameter and well polished. The remarkable thing about them, however, is that they are nearly all porphyries. The most abundant is a red feldspar-quartz porphyry. Quartzite boulders of all colors are numerous … The red porphyry boulders … have never been found in place. The source of the porphyry boulders … of this old river is a matter of great perplexity. The gravel deposit has every characteristic of an old river channel … the presence of gold in the gravels indicates their derivation from the country farther east. The river must have flowed across the gold belt, but then the question arises, how could a river of such magnitude have existed so near the summit? The only way out of this difficulty is to suppose that a great uplift has taken place along the crest and western slope, coupled with an enormous amount of erosion; and that this stream once, before this great change took place in the configuration of the country, headed many miles to the northeast, far beyond the drainage of the western slope. The boulders consist largely of hard rocks, and are very smoothly rounded and polished, indicating that they have been transported a long distance.

THE EOCENE BALLENA RIVER

Conglomerates resting on top of the Peninsular Ranges batholith in a linear outcrop pattern have been mapped by several geologists. In 1919, Arthur Ellis plotted their distribution from their northeasternmost point at Ballena Hill then trending southwest into San Diego City (*Figure 1*).

In 1935, W. J. Miller described the similarities in conglomerate-clast populations between the Ballena Gravels fluvial deposits and the Eocene Poway Conglomerate alluvial-fan strata and tied them together as parts of the same system.

In 1963, Harold Weber mapped the geology of San Diego County with an emphasis on mineral deposits including Ballena placer gold (*Figure 2*).

In 1972, John Minch turned in his Ph.D. dissertation to UC Riverside. It includes a contour map of the walls of the Eocene Ballena river valley; the walls are carved into the Peninsular Ranges batholith (*Figure 3*). The straight-line length of the exposed paleovalley is 13.5 miles with valley widths up to 2 miles. The paleovalley sinuosity ranges from 1.1 to 1.3.

The Eocene Ballena Gravels are well documented as the deposits of a large, long-distance river. What did the river look like? What were its flow characteristics?

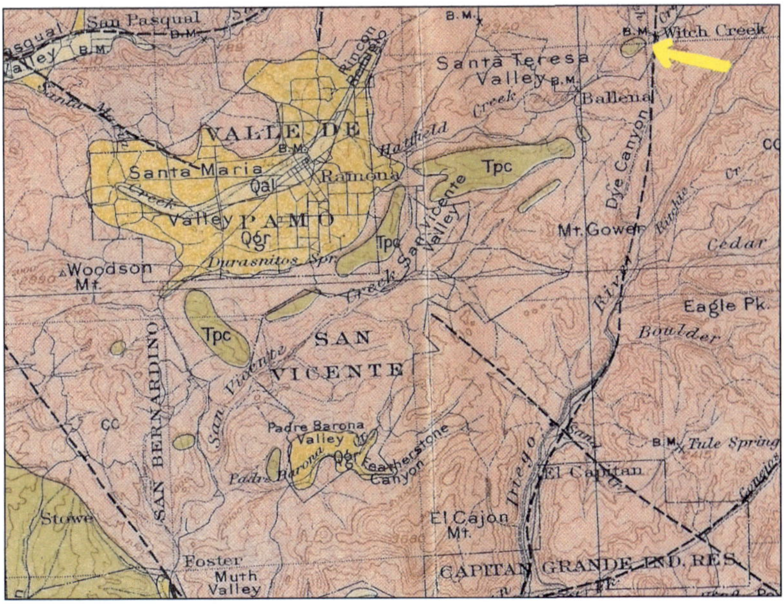

Figure 1. Portion of geologic map published in 1919 by Arthur Ellis. Near center of map, see the brown outcrop pattern labeled Tpc; this is Eocene Ballena Gravels. Extend view to upper right and see tiny brown patch that is Ballena Hill. Note the northeast to southwest outcrop pattern of Ballena River gravels.

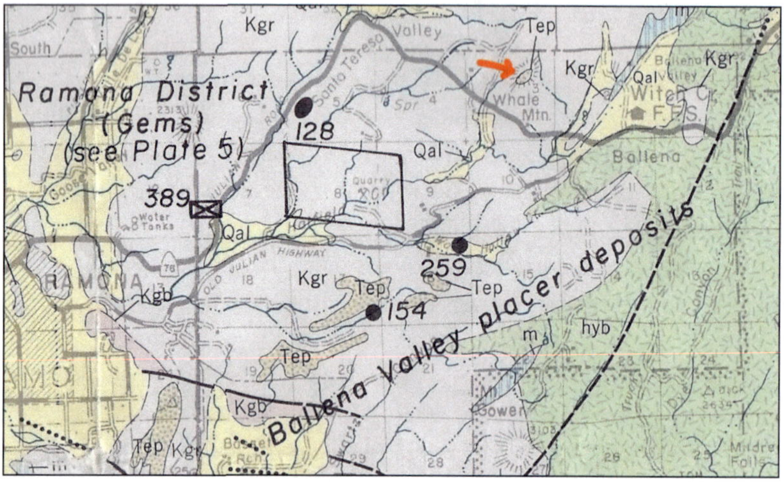

Figure 2. Portion of geologic map published in 1963 by Harold Weber. Near upper right is tiny outcrop of Ballena Gravels at "Whale Mtn." (translated to English). Also note the label "Ballena Valley placer deposits" paralleling the trend of the Eocene auriferous river gravels.

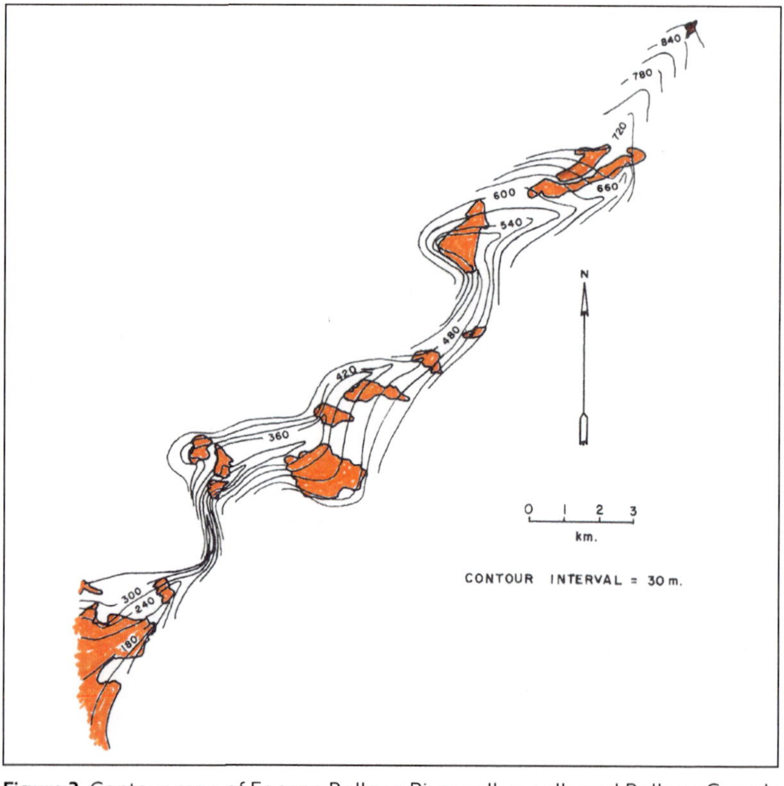

Figure 3. Contour map of Eocene Ballena River valley walls and Ballena Gravels outcrop locations by John Minch (1972). The remaining deposits of Ballena Gravels are colored orange.

PALEOHYDROLOGY OF THE EOCENE BALLENA RIVER

Paleohydrologic parameters of the gravel-bed Eocene river were assessed using a multiple-techniques approach. Equations have been derived by geomorphologists studying modern and ancient gravel-bed rivers and by civil engineers studying modern gravel-bed rivers. All of their equations were applied to describe the Ballena Gravels outcrops. Then, best-fit lines were run over all the individual results to determine the most reasonable approximation to the Eocene water flow characteristics of the Ballena River. The work was done by Brad Steer in his graduate work at SDSU aided by Professor Howard Chang of the SDSU Civil Engineering Department (Steer, 1980; Steer and Abbott, 1984).

Grain-size analyses. Special sieves were built using rebar and wood planks in order to measure grain sizes of Ballena Gravels. Each sediment sample was dug up from a one-yard square area and weighed at least 300 pounds. A butcher's meat scale was transported into the field to weigh sample fractions. The grain-size distributions were plugged into many of the hydrologic equations used in this study.

Two of the interesting sets of measurements are: 1) maximum diameter of the largest Poway rhyolite clast at a site, and 2) mean of 10 largest maximum diameters of Poway rhyolite clasts (*Figure 4*). The decrease in grain sizes to the southwest nicely defines the direction of water flow.

Using the old saying: *Coarsest closest, finest farthest* clearly indicates that the Poway rhyolite source rocks lie much farther to the east, well to the east of the Peninsular Ranges.

Gradient. The slope of the Ballena paleovalley today is about 130 feet/mile. This value is steeper than when the river flowed because of tectonic tilting to the west since Eocene time. The calculations used herein assume a downstream decrease in stream gradient with a concave-upward profile. Also used are estimates of stream length (about 200 miles) and fall distance (8,000 to 12,000 feet). The best fit estimate of Ballena River gradient in San Diego during Eocene time is 60 to 95 feet/mile.

BEDROCK SOURCE OF BALLENA GRAVELS

Where is the bedrock source of the large clasts of rhyolite and dacite in the Ballena Gravels? In 1893, H.W. Fairbanks wrote that these rocks "have never been found in place" and that they came from "many miles to the northeast, far beyond the drainage of the western slope".

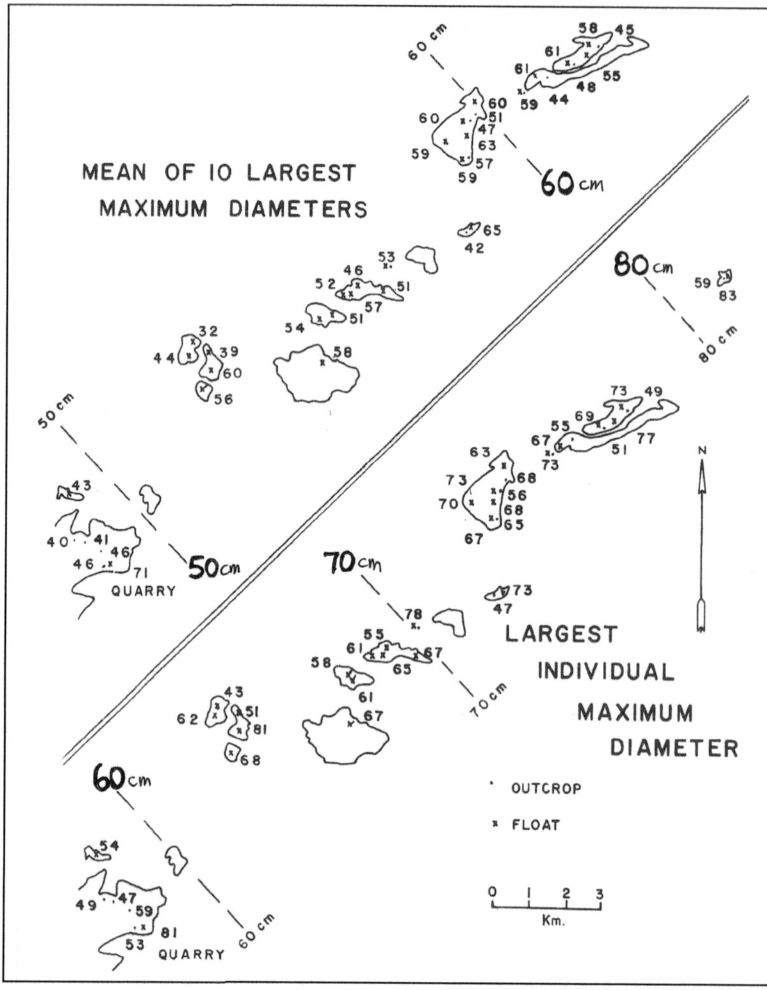

Figure 4. Plots of downstream changes in size of maximum diameters of Poway rhyolitic clasts (Steer and Abbott, 1984). At Ballena Hill, Poway clasts still exceed 32 inches diameter even after traveling for about 200 miles. The ultra-durability of Poway rhyolitic clasts allowed them to grind down the other rock types in the Ballena River and, at the same time, slow down their own attrition rate (Abbott and Peterson, 1978).

Table 1. Paleohydrologic Characteristics of the Ballena River

Sediment size	Coarse gravel
Mean grain size	1.5 to 2.2 inches
Lithology of clasts	Dominantly rhyolites and dacites

River length	About 200 miles
Drainage area	About 10,000 square miles
Sinuosity	1.1 to 1.3
Elevation of source area	7,000 to 13,000 feet
Discharge maximum	>14,000 cubic meters/second
Discharge minimum	0
Velocity typical range	5.5 to 11 miles per hour
Velocity maximum	17 miles per hour
Flow conditions	Highly variable
Flow regimes	Upper and lower
Climate	Semi-arid
Precipitation	20 to 30 inches per year
Temperature	64 to 68 degrees F annual average

In 1968, Richard Merriam suggested that his geologic map area in northwest Sonora, Mexico may contain the source rocks (his pre-Cretaceous meta-rhyolite) for the Poway Conglomerate (Ballena Gravels).

In 1972, John Minch searched through San Diego, Baja California, and northwest Sonora, Mexico working to reconstruct the geologic history of the Late Mesozoic and Early Tertiary continental sedimentation.

Definitive evidence tying the rhyolitic clasts to a specific bedrock source in Sonora came from statistical comparison of trace-element geochemistry of Sonoran Jurassic rhyolitic bedrock and rhyolitic clasts in Eocene conglomerate in San Diego (Abbott and Smith, 1978, 1989). Rhyolitic bedrock east of Tajitos, Sonora matched conglomerate clasts fed by the Ballena River to the apex of the Poway alluvial fan (*Figure* 5). Late Paleocene and Eocene rivers flowed westward from Sonora to San Diego and northern Baja California *before* the Gulf of California existed.

RECOGNITION OF POWAY RHYOLITE CLASTS

Poway clasts have several unique, distinguishing characteristics (*Figure* 6).

- Invariable presence of quartz phenocrysts
- High percentage of phenocrysts (commonly greater than 50% of rock)
- Similar sizes of phenocrysts in a rock
- No other textural features–just phenocrysts in groundmass
- Phenocrysts are sharply defined crystals that have distinct contacts with groundmass

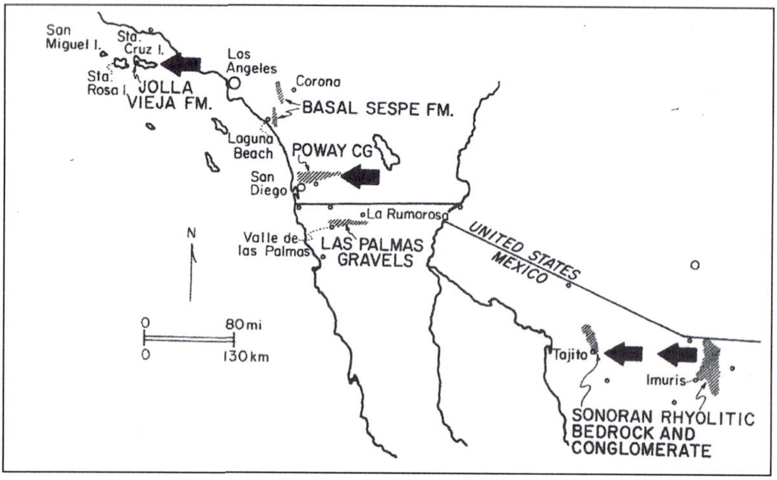

Figure 5. An integrated, east-to-west, Eocene depositional system began with erosion of Jurassic age rhyolitic-dacitic volcanic bedrock yielding clasts carried westward by Eocene rivers. The Eocene rivers flowed across the Peninsular Ranges batholith depositing fluvial sediments such as the Ballena Gravels and the Las Palmas Gravels. The Ballena River fed the massive Poway Conglomerate, an alluvial fan that buried much of San Diego. These exotic gravels then flowed westward across the shoreline, through a submarine canyon, and finally built a submarine fan, parts of which crop out on the northern Channel Islands (Abbott and Smith, 1978, 1989).

Figure 6. Poway rhyolite clast. Note: 1) the reddish color, 2) the abundance of quartz and feldspar phenocrysts, 3) the sharp contacts of phenocrysts with groundmass, and 4) absence of a preferential fabric. Photo by Pat Abbott.

- Groundmass is homogeneous and saccharoidal
- Colors mostly red through purple hues with Munsell chromas about 2; some gray clasts are commonly present
- Extreme durability of clasts makes them ring and ring as you strike them with your hammer

TWO LOCAL EOCENE RIVERS

Establishing the link between the volcanic-rock clasts in the Ballena Gravels and the Jurassic rhyolitic bedrock in Sonora allows estimation of Ballena River length. Adjusting for post-Eocene tectonic extension yields a length in excess of 200 miles. Also the Ballena River did not flow alone in Eocene time; immediately south lay the parallel Las Palmas River marked by its even thicker and more extensive conglomerates (*Figure 7*).

The Las Palmas Gravels crop out discontinuously for about 40 miles. They are mostly conglomerate filling a paleovalley up to 4 miles wide with remaining conglomerate deposits up to 785 feet thick. Its paleohydrology was assessed by Scott McDonough using a multiple-techniques approach during his SDSU

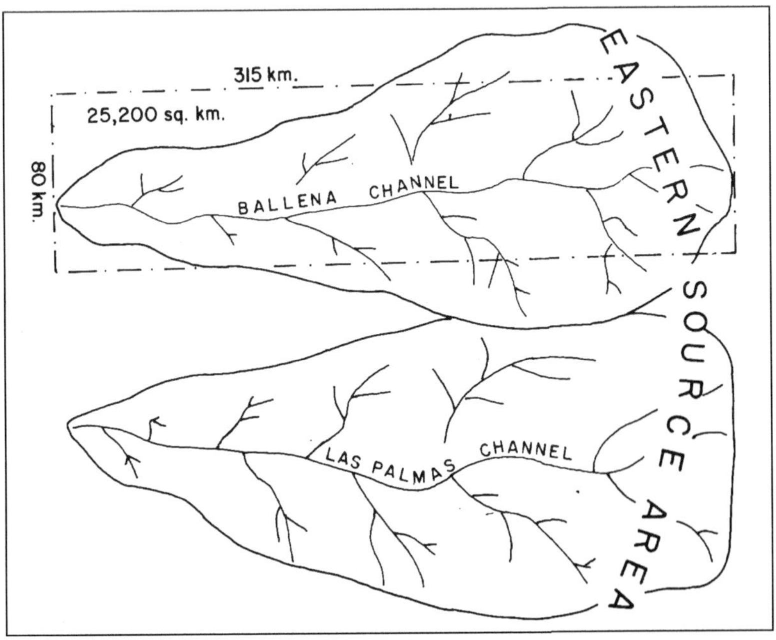

Figure 7. Relative sizes of Eocene Ballena and Las Palmas river valleys. These side-by-side river systems were each at least 200 miles long and 50 miles wide (Steer and Abbott, 1984). These Eocene rivers are analogous to the Eocene rivers that cut across the Sierra Nevada.

graduate school work (McDonough and Abbott, 1989). The Las Palmas River flow characteristics are all somewhat larger than those of the Ballena River.

The Las Palmas Gravels also are dominated by extra-regional volcanic rock clasts. Despite flowing alongside, and at the same time, as the Ballena River there are distinct differences in the mineralogical and textural compositions between the Ballena and the Las Palmas volcanic-rock clasts. The Poway rhyodacite clasts make up about 55% of the Ballena Gravels, but only about 1% of the Las Palmas Gravels (Abbott, Kies and Kerr, 1984). The Las Palmas Gravels have several unique and distinctive volcanic-rock types that make up about 58% of their clast assemblage, but they make up only about 7% of the Ballena Gravels.

REFERENCES CITED

Abbott, Patrick L., Kies, Ronald P. and Kerr, Dennis R., 1984, Right-slip offset of the Eocene Ballena river valley across the Elsinore fault zone, southern California: Ciencias Marinas, v. 9, p. 87–94.

Abbott, Patrick L. and Peterson, Gary L., 1978, Effects of abrasion durability on conglomerate clast populations: examples from Cretaceous and Eocene conglomerates of the San Diego area, California: Journal of Sedimentary Petrology, v. 48, p. 31–42.

Abbott, Patrick L. and Smith, Terry E., 1978, Trace-element comparison of clasts in Eocene conglomerates, southwestern California and northwestern Mexico: Journal of Geology, v. 86, p. 753–762.

Abbott, Patrick L. and Smith, T.E., 1989, Sonora, Mexico, source for the Eocene Poway Conglomerate of southern California: Geology, v. 17, p. 329–332.

Ellis, Arthur J., 1919, Physiography, *in* Ellis, A.J. and Lee, C.H., Geology and Ground Waters of the Western Part of San Diego County, California: U.S. Geological Survey Water-Supply Paper 446, pp. 1–50.

Fairbanks, H.W., 1893, Geology of San Diego County: California State Min. Bureau, 11th Annual Report, p. 76–120.

Henry, Christopher D., 2009, Uplift of the Sierra Nevada, California: Geology, v. 37, p. 575–576.

Kies, Ronald P. and Abbott, Patrick L., 1983, Rhyolite clast populations and tectonics in the California Continental Borderland: Journal Sedimentary Petrology, v. 53, p. 461–475.

McDonough, Scott D. and Abbott, Patrick L., 1989, Paleohydrology of the Eocene Las Palmas Gravels, Baja California, Mexico, *in*, Geologic Studies in Baja California: Pacific Section SEPM, v. 63, p. 47–62.

Merriam, R.H., 1968, Geologic reconnaissance of northwest Sonora, *in* Dickinson, W.R. and Grantz, A., eds., Proceedings of conference on geologic problems of San Andreas fault system: Stanford University Publications Geological Sciences, v. 11, p. 287

Miller, W.J., 1935, Geomorphology of the southern Peninsular Ranges of California: Geological Society America Bulletin, v. 46, p. 1535–1562.

Minch, John A., 1972, The Late Mesozoic–Early Tertiary framework of continental sedimentation, northern Peninsular Ranges, Baja California, Mexico: Ph.D. dissertation, University California, Riverside, 192 p.

Steer, Bradley L., 1980, Paleohydrology of the Eocene Ballena Gravels, San Diego County, California: M.S. thesis, San Diego State University, 147 p.

Steer, Bradley L. and Abbott, Patrick L., 1984, Paleohydrology of the Eocene Ballena Gravels, San Diego County, California: Sedimentary Geology, v. 38, p. 181–216.

Weber, F. Harold, Jr., 1963, Geology and Mineral Resources of San Diego County: California Division Mines and Geology, County Report 3, 309 p.

Origin of "Pseudo-Tsunami" Deposits in the Ramona Area, San Diego County, California

WILLIAM J. ELLIOTT *(Solana Beach, CA)*

ROY J. SHLEMON *(Newport Beach, CA)*

ABSTRACT

Patches of exhumed weathered granitic bedrock are common in the Ramona area of southern California. Following a San Diego Association of Geologists "One-Stop-Wonder" field trip on October 20, 2012, we visited Sites 1, 2 and 3 on January 12, 2013. The senior author (WJE) examined Sites 4 and 5 on January 16, 2013. Our specific focus was to assess whether or not features exposed at these sites were indeed paleo-tsunami deposits as postulated by Busch and others (2012). After evaluating reasonable alternative hypotheses, we conclude that features exposed within these small isolated outcrops are, in fact, remnant paleosols, weathering products of an extended period of Pleistocene landscape stability. We suggest that the "paleosol hypothesis" is more reasonable than a presumed artifact of the ~2.51 Ma Eltanin bolide impacting the southern Pacific Ocean and producing a tsunami that swept across the Pacific leaving "trace" sediment evidence as far inland as Ramona.

INTRODUCTION

The 2004 Indian Ocean and the 2011 Tohoku-oki (Fukushima), Japan, tsunamis caused extensive damage and loss of life, and also unleashed new interest in identifying possible paleo-tsunami deposits in California generally, and in southern California in particular. Of special interest are Pacific Ocean "mega-tsunamis" generated by asteroids that may have caused ocean-wide impact.

To date, the only known asteroid impact in a deep ocean basin is the Plio-Pleistocene Eltanin event that occurred in the eastern southern Pacific Ocean. Evidence for this bolide was first recognized in 1981 as an iridium anomaly in sediment cores taken by the research vessel USNS Eltanin in 1965. The Eltanin impactor left behind an approximately 80,000 km^2 chaotic debris field that included ripped up Eocene and possibly Paleocene sediments (Gersonde and others, 2003). The impact site is located at 57.78°S, 90.79°W, about 715 miles (1153 km) southwest of Isla Carlos on the southwestern side of the southern tip of Chile (Ward and Asphaug, 2002, p. 1076, Google Earth, 2017).

Later investigations (www-1, 2017; Goff and others, 2012; and references therein) identified a high iridium area that suggested an estimated ~0.6 to 2.5 miles (1 to 4 km) wide asteroid may have triggered a widespread mega-tsunami. Magneto-stratigraphy and microfossils were used to date the Eltanin bolide at ~2.51 Ma, a time generally coinciding with the Plio-Pleistocene boundary and climatic transition (Gersonde and others, 2003).

Though fraught with uncertainty, numeric modeling suggests that initial (T+1-hour) wave heights from the Eltanin impact may have ranged from about 846′ to 942′ (258 to 287 m) in the open ocean, but were greatly attenuated with distance (T+10-hours) from the point source to less than ~115′ to 131′ (~35 to 40 m) as the sea wave approached the Central American coast (Ward and Asphaug, 2002; see also, Goff and others, 2012; and Weiss and others, 2015).

Recently, Busch and others (2012) inferred that several "unusual sedimentary deposits" in San Diego County could be manifestations of the Eltanin impact. Because of an absence of accurate location data in their original publication (first printing), Busch and others led local geologists to Sites 1, 2 and 3 (*Figure 1*) on October 20, 2012, for a personal look at their "ancient tsunami deposits." Accordingly, based on our follow-up observations and measurements, we agree with Busch and others (2012, p. 66) who astutely pointed out that "the tsunami interpretation is considered speculative" and will require further testing before general acceptance can be achieved. We now provide "testing" from observations and measurements at five sites and, accordingly, reject the "tsunami hypothesis" in favor of a "paleosol origin" – a far more rational explanation.

As pointed out by the late Carl Sagan "extraordinary claims require extraordinary evidence" (Sagan, 1995; Pinter and Ishman, 2008). We therefore specifically examined three of the several Busch and others (2102) purported Eltanin tsunami deposits in the Ramona area of San Diego County. We particularly focus on the geomorphic setting, on the character of sediments within and surrounding the inferred tsunami deposits, and on the soil-stratigraphic context. Additionally we address elevation differences and ideas regarding isostatic/eustatic changes. Accordingly, we now provide an alternative origin for the features; one based on soil (pedogenic) weathering processes.

The senior author (WJE) examined two additional Sites, 4 and 5 (*Figure 1*) on January 16, 2013. These sites are similar in character to Sites 1, 2, and 3, and are included in *Table 1* to provide additional data and technical observations for the Ramona region. In this regard, we find that the "floating" boulder shown by Bush and others in their *Figure 8* (2012, p. 75) is nothing more than a displaced, errant core-stone (weathered tonalite boulder) that tumbled down onto an existing exposed pedogenic surface (*Photo 5*).

We also note that Busch and others (2012) identified other localities in San Diego County and elsewhere in southern California as possible examples of Eltanin mega-tsunami deposits. We have not specifically examined those sites, and hence confine our comments to the Ramona area (*Figure 1*).

STRATIGRAPHIC CONTEXT

A paleo-tsunami is one that occurred prior to the historical record or for which there are no written records of firsthand observations (Goff and others, 2012). Accordingly, such tsunamis are generally inferred by their proxy records, namely: the thickness of fining-upward and laminar sand sheets, possible rip-up clasts near the base or within a particular deposit and, most telling, the presence of large, beach-derived cobbles and boulders on top of sandy units. But, distinguishing between deposits laid down by tsunamis and by high storms surges is not easy, despite many competing criteria put forth by several authors (for example: Cantalamessa and others, 2005; Morton and others, 2007; Bridge, 2008; Bahlburg and others, 2010).

Caution is therefore clearly required. Indeed, as pointed out by Goff and others (2012), the lack of tsunami proxy records (for example, foraminifera and chaotic deposits) should not be taken as evidence to refute the purported tsunami origin; but too, the presence of only a few proxies should not, on their own, be construed as evidence for their presence.

A classic example of misinterpretation comes from chaotic deposits along the northern Chilean coast where Holocene (and historic) tsunami deposits are to be expected given the tectonic framework of the area. Here, Cantalamesa and Di Celma (2005) described intermixed boulders and sand, rip-up clasts and intercalated fossil beds and construed these to be paleo-tsunami deposits. However Bahlburg and others (2010) revisited the site, made their own measurements and descriptions and concluded in no uncertain terms that the deposits are ".... subaqueous equivalents of graben-flank debris-flow, alluvial-fan breccias, and not tsunami [debris] of any kind" (p. 80).

The challenge of identifying and dating paleo-tsunami deposits is not limited to the east coast of the Pacific. For example, boulders 98 feet (30 m) above modern sea level on the southeastern Australian coast were long thought to be an artifact of a Holocene tsunami; now, however, these "high-level" deposits are believed to record evidence of the ~2.51 Ma Eltanin event (Goff and others, 2012). Accordingly, as in much of geologic interpretation, some "strange" deposits are thought to record paleo-tsunami events, but the origin of others is still equivocal. There is certainly no shortage of Multiple Working Hypotheses (Chamberlain, 1890) all eventually subject to rejection or tentative acceptance.

"PSEUDO"-TSUNAMI DEPOSITS IN THE RAMONA AREA

We examined three Ramona-sites on January 12, 2013, interpreted by Busch and others (2012) as evidence for ~2.51 Ma Eltanin tsunami deposits (See *Figure 1* and *Table 1* for location and outcrop data).

> **Site 1** is exposed in an unnamed, southwest-trending arroyo ~1.5 miles (2.5 km) northeast of Bassett Ranch, and ~1.3 miles (2.1 km) west of the Mount Gower Open Space Preserve, (USGS Ramona 15′ topographic quadrangle map, 1947; USGS Ramona 7.5′ topographic quadrangle map, 1955).
>
> **Site 2** is in an arroyo cut about 280′ (85 m) upstream, on the opposite bank from Site 1 (see Busch and others, 2012, 2nd printing, *Figure 5* [photograph] on p. 74).
>
> **Site 3** is in road cuts on both sides of the Old Julian Highway, ~1.35 miles (2.1 km) west of the "Y" intersection of Old Julian Highway and SR-78 (former community of Ballena), here informally designated as the "Power Pole Site" after a power pole on the south side of the road.

At Sites 1 and 2, Busch and others (2012) interpreted an unnamed muddy sandstone unit (mapped as QTs in the Ramona and San Pasqual 7.5′ quadrangles; Todd and others, 2006; Hernandez and others, 2007) as likely having one of three possible origins: (1) local flash flood deposits; (2) late-stage flood deposits of the Tertiary Ballena River system; and (3) indurated remnants of a paleo-tsunami sheet flood and return flow (p. 65, 66). Busch and others (2012) also point out that: "Due to the elevation and distance to the coast, the tsunami [Eltanin ~2.51 Ma event] interpretation is considered speculative…." (p. 66). Based on our observations and as pointed out in the following discussion, we wholeheartedly agree that the paleo-tsunami interpretation is indeed very speculative.

Busch and others (2012) used two main lines of evidence to postulate Eltanin tsunami deposits in the Ramona area:

1. Observation of flood deposits that contain angular fragments of locally derived pegmatitic rock, rare "Poway-type" metavolcanic and quartzite clasts (derived from nearby Ballena gravels), and a few "rotten" granitic clasts in a sandstone-supported matrix (p. 65); and
2. Lowering the Peninsular Ranges ~1° around an axis located near the present-day coast line to place Sites 1, 2, and 3 at or near sea level during the ~2.51 Ma Eltanin bolide.

We test the paleo-tsunami speculation by reconstructing the general geomorphic setting of Sites 1, 2, and 3 by: (1) soil-stratigraphic measurement and description at Site 2; (2) evaluating the significance of elevation differences between Sites 1, 2, and Site 3; and (3) assessing the timing and impact of regional isostatic/eustatic changes.

GEOMORPHIC SETTING

All three sites are underlain by quartz diorite (tonalite) and plutonic rocks of the mid- to late-Cretaceous Peninsular Ranges batholiths (Weber, 1963; Rogers, 1966; Walawender, 2012; Todd and others, 2012).

Immediately north of Sites 1 and 2 (*Figure 1*), an east-west trending ridge-line is capped by remnant gravels of the Eocene (~44 Ma) Ballena fluvial system ("Poway-type clasts"). Today, these gravels are preserved as a topographic inversion of topography (Abbott, 1999, p. 86–93). This unnamed ridge-line rises ~ 985–1640′ (~300–500 m) above paleosols in the arroyo cut into the valley below (*Figure 1*).

For at least several million years, eroded Ballena gravels have moved, and continue to move downslope, lap onto, and interfinger with interbedded sand- and silt-sized sediments that comprise Sites 1 and 2 "fan/fluvial" deposits. These "fan/fluvial" deposits fill many minor valleys that likely owe their origin to repeated cutting and filling during regional Pleistocene changes in climate and vegetation. That multiple epochs of channel incision followed by relative landscape stability is indicated by the stacked, strongly developed buried paleosols exposed in arroyo cuts at Sites 1 and 2 (Photos 1 and 2).

SITES 1 AND 2

As shown on *Figure 1*, Sites 1 and 2 are about 280′ (85 m) apart as exposed on opposite sides of a 13-ft (~4 m) deep arroyo cut. Paleosols exposed in this arroyo, both relict and buried, are weathering phenomena superimposed on underling siliclastic "fan/fluvial" sediments (relict paleosol), and weathered quartz diorite (tonalite) bedrock (buried paleosol).

Both sites expose a similar stratigraphy. From an access standpoint, we selected Site 2 (*Photo 2*) for measuring and describing a representative soil-stratigraphic section. Accordingly, as documented on *Table 2*, we identified four discrete soil horizons useful for assessing landscape evolution, and sediment (parent material) origin and age.

SOIL HORIZON 1. Surficial Soil Horizon

The uppermost horizon (A/C) is a thin medium to coarse sandy loam (*Table 2*). Generally loose and structureless, this horizon consists of a veneer of modern to possibly early Holocene "fan/fluvial" deposits derived from overland flow and in-situ weathering. A few modern roots penetrate the horizon and extend downward along ped faces within the near-surface relict paleosol (2Btb). The light brown color (Munsell 7.5YR-6/3) is locally variegated, presumably owing to periodic minor influx of eolian dust and to reworking of underlying sediments.

In conventional geologic terminology, this A/C soil horizon is equivalent to "Qc" (Quaternary colluvium). In general, it rarely exceeds about 6" (15 cm) in thickness.

SOIL HORIZON 2. Relict Paleosol

The 2Btb horizon (*Table 2*; Photos 1 and 2) is a moderately to strongly developed, relict paleosol with a thickness of up to ~6.5′ (~2 m). Typically, it is a strong brown to reddish brown (5YR-4/4–5/4) loamy fine sand with strong medium angular blocky to locally prismatic structure. Few to common reddish-brown (5YR-4/4) clay films line ped faces and bridge mineral grains.

Geologically, this horizon formed on an internally stratified, generally fining-upward sedimentary sequence. Locally, it contains scattered pockets of interbedded and reworked granitic gravel and, sub- to well-rounded ovoid Ballena-derived clasts. This soil, though often occurring at or near the modern ground surface, formed mostly under climatic and vegetative regimes of the past; it is therefore designated as a "relict paleosol."

Relative profile development indicates weathering for at least ~40 ka (marine isotope stage 3), but more likely in the ~100 ka range (stage 5).

SOIL HORIZON 3. Buried Paleosol—Silcrete

AT SITE 1. Site 1, in particular, clearly exposes a unique, < ~0.5-ft (15 cm) thick, now-buried siliceous duripan (silcrete; horizon 2Bqmb; *Table 2*). As observed at Sites 1 and 2, this pedogenic horizon is a distinctive, widespread, laterally traceable, stratigraphic marker capping underlying strongly weathered quartz diorite (tonalite) regolith (*Photo 1*).

The duripan perches modern gravitational water and is essentially impenetrable to roots, which spread out horizontally. These ~1 to 2 inch (2.5 to 5 cm) deep and ~10 to 20 inch (25 to 50 cm) wide former-root-systems are now preserved as irregularly shaped quasi-polygons typically filled with silt. For the most part, silt infillings were translocated downward along seasonally open ped faces within the overlying relict paleosol (*Table 2*). Some crack-filling silts are slightly cemented and brittle, similar to a classic pedogenic fragipan.

The silcrete polygons are extremely hard and firm, with reddish brown (5YR-5/4) to yellowish red (5YR-4/4) colors (*Photo 1*) attesting to strong to very strong pedogenic development. They are judged to represent ~100 to 300 ka of weathering.

AT SITE 2. The Site 2 silcrete (*Table 2*) is particularly important for assessing the age and origin of features cited by Busch and others (2012, p. 64–66) as evidence for the Eltanin tsunami. Specifically, silty fragipans in the quasi-polygons owe their origin to filling of original horizontal root voids; and are not tsunami rip-up clasts.

Further, the silcrete is <u>not</u> an exhumed paleosol—where overlying *sedimentary strata* have been eroded to re-expose a once buried, former weathering surface. Rather, the horizon, though of considerable age, is still accreting even though buried by the overlying relict paleosol (horizon 2Btb).

Accordingly, the silcrete (2Bqmb) horizon, identified by Bush and others (2012, p. 64–66) as containing tsunami deposits, is pedogenically late Pleistocene in age. These pseudo-tsunami deposits are, in reality, normal pedogenic weathering products and not stratigraphic evidence for an inferred ~2.51 Ma south Pacific, Eltanin tsunami event.

SOIL HORIZON 4. Buried Paleosol—Weathered Bedrock

The basal, regionally extensive, buried paleosol identified at Sites 1 and 2 (*Figure 1*, Photos 1 and 2) formed on weathered quartz diorite (tonalite) bedrock. This horizon, designated "3Btb" (*Table 2*), is a strong brown to dark brown (7.5YR-4/6) to dark brown (7.5YR- 4/4) medium to coarse loamy sand. Its moderate medium-angular blocky structure decreases to weak, fine, sub-angular blocky to massive structure near the base of the measured section (*Table 2*). It displays common fine- to medium- reddish-brown (5YR-4/4) clay films that line ped faces and bridge mineral grains. Based on its thickness and strong profile development, the 3Btb paleosol represents another ~100–300 ka of weathering.

SITE 3

Site 3 is exposed in road cuts on both sides of Old Julian Highway (*Figure 1, Table 1*), readily identified by a power pole on the south side of the road (*Photo 3*). Here we note the presence of buried paleosols (though less well developed) and *in situ* mud cracks similar to the Site 1 and 2 arroyo exposures. Grussified granitic clasts are present—attesting to periods of intense weathering. Although "negative evidence," we nowhere found stratigraphic features even vaguely suggestive of a paleo-tsunami, much less the ~2.51 Ma Eltanin event.

A particularly problematic wrinkle is that Site 3 is ~695′ (212 m) higher in elevation than Sites 1 and 2, at a distance of only ~2.8 miles (4.5 km). Mechanically, getting the hypothetical Eltanin deposits up to this elevation in such a short distance is seemingly nonsensical.

PSEUDO-TSUNAMI RUNUP

Pseudo-Tsunami Elevations: Based on wave attenuation assumptions of Ward and Asphaug (2002, p. 1076–1077), Busch and others (2012, p. 58) note that the ~2.51 Ma Eltanin event may have had a coastal impact of perhaps about 98′ (30 m) above paleo-sea level along the Southern California coastline.

Accordingly, identifying such deposits at Sites 1, 2 and 3, at elevations of 1,705′ (520 m), 1,735′ (529 m), and 2,415′(736 m), respectively, (*Table 1*), requires *post* paleo-tsunami regional uplift, of 1°, with the hinge-line axis paralleling the present coast (Busch and others, 2012, p. 66).

In combination with their assumed 1° tilt, Bush and others (2012, p. 66) therefore rely on Hertlin and Grant (1944, p. 29) to assert that a nominal Plio-Pleistocene sea level was ~800′ (~244 m) above present sea level.

The Hertlein and Grant (1944) paper is clearly a "classic;" nevertheless assumptions of paleo-sea levels some 73 years ago have long since been superseded by more definitive calculations. We therefore reject use of this outdated information used to propose a once higher sea level at ~2.51 Ma.

It is likely that Hertlin and Grant (1944) interpreted the ~800-foot (244 m) Poway terrace (*Figure 3*) to represent a sea level marker–which it may well have been during Eocene time when "Poway-type" gravels were being deposited by rivers flowing westward from Sonora, Mexico (Abbott 1999, p. 85–104). For context, the 1944 work was done well before our present-day understanding of plate tectonics and the northwestward shift of the west coast of North America and Baja California along the San Andreas fault (Atwater, 1970).

Fast forward 73 years, relatively datable coastline stratigraphic markers are now provided by Pleistocene terrace and wave-cut platforms intermittently exposed in San Diego County from Point Loma on the south to beyond Camp Pendleton on the north. The highest remnant terraces on Camp Pendleton were judged to be over a million years old (Shlemon, 1978, 1987); and those in the San Diego area were determined to be up to ~1.6 Ma (Kennedy and Tan, 2008). These emergent terrace flights were cut by glacio-eustatically controlled sea levels at or near present sea level and then uplifted with the passing of time.

The Plio-Pleistocene high sea level occurred when temperatures were ~9° to 14½° F (5° to 8° C) warmer than the present, not dissimilar from marine isotope stage 5e (~125 ka) when the then interglacial sea level was ~19.7′ (6 m) higher than the present (See, for example, Melles and others, 2012, Brigham-Grette and others, 2013).

Accordingly, for Eltanin deposits to be preserved at Sites 1, 2 and 3 requires coastal wave impact many tens of meters above the likely Plio-Pleistocene shoreline, and with seemingly unbelievably high velocities. This is particularly apparent when accounting for the presumed 695′ (212 m) elevation run-up between Sites 1, 2 and Site 3, across a distance of only 2.8 miles (4.5 km) (Figures 2 and 3, *Table 1*).

Many estimates have been proffered about the amplitude of paleo-sea level fluctuations during the past several millions of years (see, for example, Haq and others, 1987; Wardlaw and Quinn, 1991; Kominz, 2001, Miller, 2009, and references therein). Besides constraints to the magnitude of sea levels previously indicated, Wardlaw and Quinn (1991) conclude from studies at Enewetak Atoll that Pliocene sea-levels did not rise higher than 118′ (36 m) above, or 207′ (63 m) below present sea level. Also, during Pleistocene-age glaciations large volumes of water are typically sequestered in polar and continental ice sheets, resulting in sea levels nominally lower than today. It is difficult to ignore empirical science while arguing for a much higher sea level at the time of the Eltanin bolide as a means for sea level rise to the elevation of Ramona at the Plio-Pleistocene boundary. We therefore conclude that such a feat is, at best, speculative and unsupportable.

Likelihood of general landscape tilt since the ~2.51 Ma Eltanin event. One degree east-side-down rotation of the Peninsular Ranges (Busch and others, 2012, p. 66), around an axis located about 20 miles (32 km) west from Sites 1 and 2 (immediately east of the eastern edge of the 400 foot (122 m) Linda Vista terrace, Figures 2 and 3), would lower Sites 1 and 2 about 1,805′ (550 m) at ~2.51 Ma. This would place Sites 1 and 2 about 85′ (26 m) below sea level and Site 3 about 610′ (186 m) above sea level.

Site 3, would then be stranded about 695′ (212 m) above Sites 1 and 2, and more than likely beyond the reach of a tsunami wave (see Calculation Module). To address this problem, we suggest that Site 3 may have been at or near the same elevation as Sites 1 and 2 at the time of the tsunami. Then, some time since ~2.51 Ma, vertical faulting elevated Site 3 to its present elevation–a scenario not supported by recent mapping by Todd and others (2012).

In contrast, common sense suggests that it is *unlikely* the Peninsular Ranges just *dropped down* at ~2.51 Ma—long enough for the Eltanin tsunami to occur—and then just *popped back up* to the present elevation. It seems that a 1° lower elevation would have to have been in place for some measurable length of geologic time–during which time the Eltanin tsunami occurred. If this were the case, then some remnant of former marine conditions should still be present between Sites 1, 2 and 3 and the coast-line; for example, (1) fossil-bearing marine sediments; (2) strath terraces along canyon and valley sides between Sites 1, 2 and 3, and the coast, and (3) knickpoints cut into canyon and valley sides at the backs of strath terraces. To our knowledge, this type of geologic evidence has never been observed or reported for the terrain between the coast and Sites 1, 2 and 3.

Potential energy between Sites 1, 2 and 3. A tsunami wave, when approaching shore, shoals (water depth decreases)—with a resulting wave amplitude increase and forward velocity decrease. Additionally, as the tsunami wave travels inland, beyond the sea shore, it expends kinetic energy traveling up slope and carrying entrained debris collected along the way. Eventually the wave runs out of energy to go further inland and begins its retreat to the sea (Ward and Asphaug, 2002, Paine, 2017, www-2, 2017).

It is one thing to get the Eltanin tsunami from the coast to Sites 1 and 2 (perhaps by the aforementioned ~1° tilt, or by a much higher regional mean sea level at ~2.51 Ma). However getting that same tsunami run-up wave to continue its climb up from Sites 1 and 2 to Site 3 is problematic. To accomplish this, the sea wave would have to have had a residual velocity of ~144.5 miles/hour (232.5 km/hour) to lift one cubic foot (0.028 m^3) of sea water up 695′ (212 m) to Site 3.

To reach Sites 1 and 2, the sea wave would have to have traveled ~15 to 20 miles (24 to 32 km) inland, threading its way through tortuous canyons and valleys to reach the Ramona area. Assuming such a sea wave could actually reach Sites 1 and 2, it seems highly unlikely that sufficient kinetic energy (velocity mass) would remain to continue driving the wave up another 695′ (212 m) to Site 3 (see the Calculation Module and Figures 2 and 3).

CALCULATION MODULE

The logic follows that of Sears (1950–58, p. 124–25)

- Potential Energy (PE) = Kinetic Energy (KE) (also known as Conservation of Energy)
- PE = mgh (mass gravity height)
- KE = ½mv^2 (m = mass, v = velocity)
- g = gravity = 9.8 meters per $second^2$
- For a height, h, of 212m (695 feet), and a mass of one cubic foot of sea water, m, 29.0 kg (64.0 pounds) :
  PE = 29.0 kg 9.8m/s^2 212m = 60,250 Joules of potential energy
- PE can also be represented as: PE = ½mv^2 (m = mass, v = velocity)
- Substituting: 60,250 Joules = ½ 29.0 kg v^2
- Solving for v: v = the square root of 2 × 60,250 divided by 29.0 = 212 feet/second (232.5 km/hour or 144.5 miles/hour)

Therefore, the sea wave would have had to have a velocity of ~144.5 mi/hr (232.5 km/hr) when reaching Sites 1 and 2 in order to continue on up another 695 feet (213 m) in elevation, over a distance of ~2.8 miles (4.5 km), and up a tortuous slope of 2.71°, to reach Site 3. A seemingly improbable task.

If Site 3, rather than Sites 1 and 2, were at or near sea level during Plio-Pleistocene time (~2.51 Ma), they would have been too far below sea level to have received hypothesized tsunami deposits.

To account for the 695′ (212 m) elevation difference between Sites 1, 2 and 3, one must call on heretofore unrecognized intervening faulting or folding to be present, a scenario not supported by field observations or by recent geologic mapping (Todd and others, 2012).

SUMMARY AND CONCLUSIONS

The 2004 Indian Ocean and the 2011 Tohoku-oki (Fukushima), Japan tsunamis renewed interest in identifying possible evidence for pre-historic tsunamis in southern California. A possible candidate is the Eltanin impactor, likely an asteroid that plunged into the Southern Pacific Ocean, 715 miles (1153 km) southwest of Isla Carlos on the southwestern side of the southern tip of Chile. Dated by magneto-stratigraphy and microfossils, the ~2.51 Ma Eltanin event is calculated to have created T+1-hour tsunami wave heights of up to ~920′ (280 m) near the point source in open-ocean. By T+10-hours, tsunami wave heights closing in on the coast of Central America are estimated to have been reduced to ~115′ to 131′ (35 to 40 m) (Ward and Asphaug, 2002, p. 1076, 1077). At the southern California coastline, estimated wave heights are "guestimated" to have been about ~98′ (30 m). The impact age corresponds to regional climate change, generally coincident with the Plio-Pleistocene boundary.

Busch and others (2012) speculate that several remnants of Eltanin deposits are preserved in southern California; we examined three of their proposed sites (Sites 1, 2, and 3) in the Ramona area. Our field examination and soil-stratigraphic documentation of these "pseudo-tsunami" deposits shows that an upper relict paleosol is exposed in cuts at Sites 1 and 2, where classic, climatically controlled, fining-upward (basal gravels and coarse- to fine-grained sand and silt sized particles) fluvial ("flood") deposits are exposed. These sediments contain vertical fossil roots as well as a moderately to strongly developed solum.

A strongly developed buried paleosol, below the above-noted relict paleosol deposits, bears a distinctive and locally resistant siliceous duripan (silcrete) that acts as a barrier to modern gravitational water. These crudely polygonal clayey deposits preserve paleo-root systems now filled with relatively resistant silt, now forming vertical, 0.4" to 0.8" (1 to 2 cm) high fragipans.

These features are typical of seasonally wetted pedogenic clay within drylands so characteristic in the western United States. In essence, we find that the inferred tsunami rip-up deposits and clasts within the sand and silt matrix are not Eltanin, ~2.51 Ma paleo-tsunami deposits, but rather the products of normal pedogenesis that has occurred within the last few hundred thousand years.

In addition to pedogenic evidence, we also reject the tsunami origin based: (1) on elevations of, and elevation differences between, sites 1, 2 and 3, (2) on their distance from the modern coast, (3) on assumed tilt rates, and (4) on Plio-Pleistocene sea levels needed to transport Eltanin waves to the elevations of: Site 1 (1,705 feet, 520 meters), Site 2 (1,735 feet, 529 meters) in general, and to Site 3 (2,415 feet, 736 meters) in particular.

Compelling evidence for Eltanin paleo-tsunami event may eventually be found; but the five Ramona-area deposits are more easily and better explained by normal fluvial sedimentation and by pedogenic processes. In accordance with common sense and with the dictum that "extraordinary claims [~2.51 Ma Eltanin tsunami deposits in San Diego County] require extraordinary proof," we therefore reject the tsunami claim as indeed speculative.

ACKNOWLEDGMENTS

We thank Chris Livesey for encouraging us to complete this long overdue project. Larry Busch graciously invited discussion and provided us with a complimentary copy of the Preliminary CGS Geologic Map of the Ramona 7.5′ Quadrangle. We likewise thank Dr. Elaine Hanford for stimulating discussions and for assistance in obtaining pertinent publications. Dr. Monte Marshall provided useful insights into the physics of kinetic and potential energy, and with tongue in cheek, observed that: *"Oh! so it ain't Beach Sand ?"* And of course, we thank Wendy Elliott for her patience and encouragement — always wondering "are we there yet?"

ILLUSTRATIONS

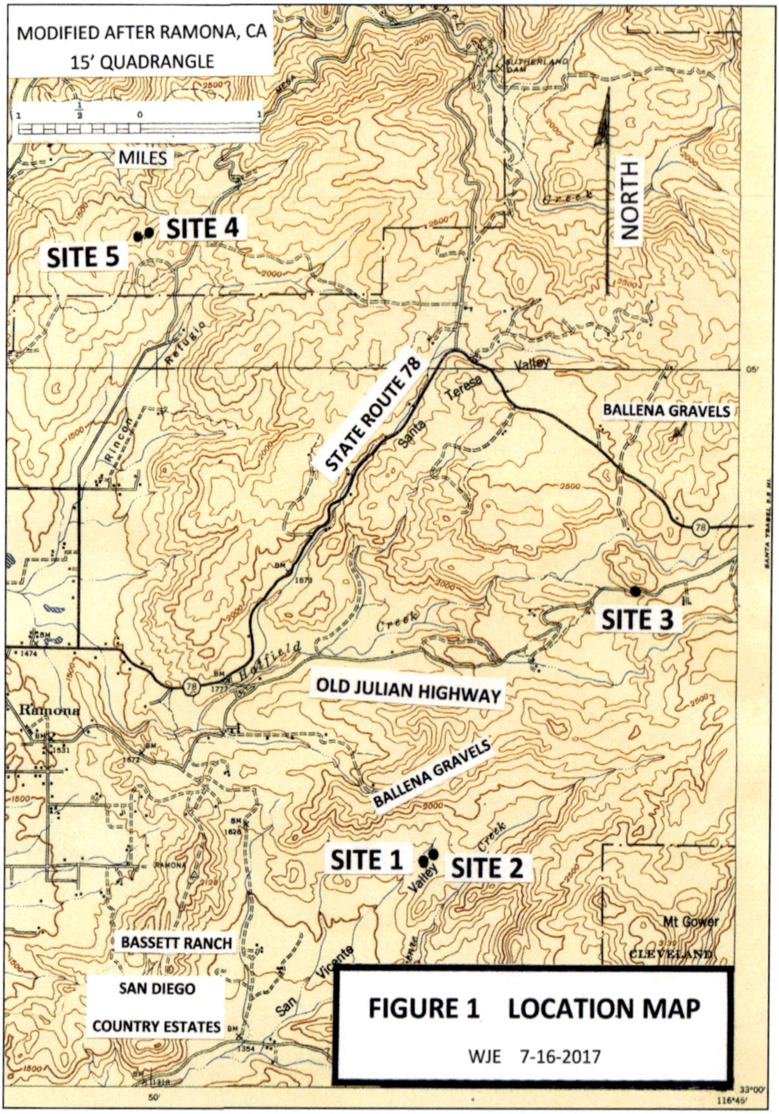

Figure 1 Location Map, Sites 1, 2, 3, 4, and 5.

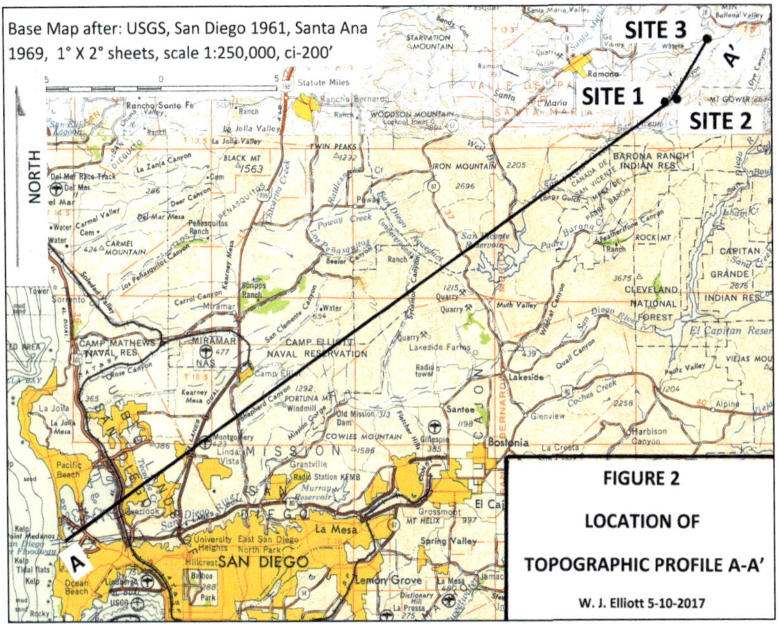

Figure 2 Location Map, Showing Location of Figure 3—Topographic Profile.

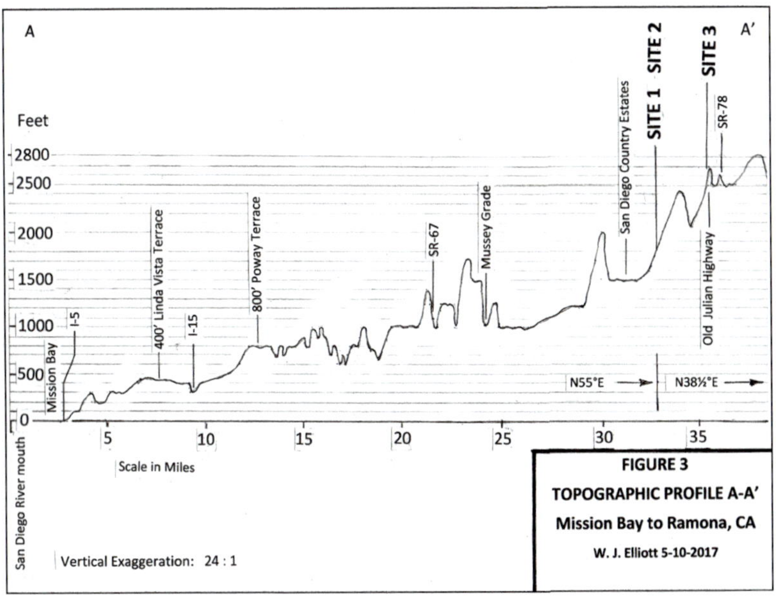

Figure 3 Topographic Profile, Proposed Run-up from the ~2.51 Ma Eltanin Bolide.

Table 1 Outcrop Location Data.

OUTCROP LOCATION DATA (*See* Figure 1 *for Site locations*)						
SITE	**LAT.**[1]	**LONG.**[1]	**SECTION**[2]	**TOWNSHIP**[2]	**RANGE**[2]	**EL.**[1]
1	33° 01′ 35.16″ N	116° 47′ 42.74″ W	SE¼, NW¼, SE¼, Section 20	T13S	R2E, SBBM	1,705′ 520 m
2 (*Bush, 2012, 2nd printing, Fig. 5, p. 74*)	33° 01′ 35.69″ N	116° 47′ 39.08″ W	SE¼, NW¼, SE¼, Section 20	T13S	R2E, SBBM	1,735′ 529 m
3	33° 03′ 28.28″ N	116° 46′ 01.17″ W	N ½, SW ¼, Section 10	T13S	R2E, SBBM	2,415′ 736 m
4 (*Bush, 2012, 2nd printing, Fig. 7, p. 75*)[3]	33° 05′ 56.93″ N	116° 50′ 12.07 W	SW¼, NE¼, SW¼, Section 25	T12S	R1E, SBBM	1,710′ 521 m
5 (*Bush, 2012, 2nd printing., Fig. 8, p. 75*)[3]	33° 05′ 56.44″ N	116° 50′ 12.85″ W	SE¼, NW¼, SW¼, Section 25	T12S	R1E, SBBM	1,713′ 522 m

The distance between Sites 1, 2 and Site 3 is 2.8 mi. (4.5 km)

[1] Obtained online from Google Earth.

[2] Obtained from United States Geological Survey 1955.

[3] Sites illustrated by Bush and others (2012, second printing, Figures 7 and 8, p. 75).

These were visited by WJE and included here only for sake of completeness.

Table 2 Generalized Soil-Stratigraphic Section, Ramona, California.

GENERALIZED SOIL-STRATIGRAPHIC SECTION, RAMONA, CALIFORNIA (*See* Table 1 *for location, and* Photos 1 *and* 2)			
Bush and others, 2012, 2nd ed.[5]	**DEPTH FEET**	**PEDOGENIC SOIL HORIZON**	**DESCRIPTION**
Qc	0.0–0.3	A/C	Light brown (7.YR-6/3) when dry medium to coarse sandy loam, structureless; hard, friable, non-sticky and non-plastic; few modern roots; abrupt smooth boundary.
	0.3–6.5	2Btb	**Relict Paleosol:** Strong brown (7.5YR 4/6) to reddish-brown (5YR 5/4) loamy fine sand; very strong angular blocky to locally prismatic structure; hard to very hard; locally friable; non-sticky and non-plastic; few to common reddish brown (5YR 5/4–5YR-4/4) clay films bridging mineral grains and locally lining ped faces; abrupt smooth to wavy boundary. Geologically, the sediments represent a locally internally stratified, generally fining-upward section with interbedded, weathered and reworked granitic gravel and oviod Ballena-derived clasts. The 2Btb horizon is moderately to strongly developed, representing ~40–100 ka of *in situ* weathering (see Figure 1, Photo 2 labeled "RELICT PALEOSOL").

GENERALIZED SOIL-STRATIGRAPHIC SECTION, RAMONA, CALIFORNIA (*See* Table 1 *for location, and* Photos 1 *and* 2)			
Qts	6.5–7.0	2Bqmb	**Buried Paleosol:** Undifferentiated, reddish brown (5YR-5/4) to yellowish red (5YR-4/4) siliceous duripan (2Bqmb); locally brecciated with ~1–2 inch deep and 1-in wide quasi-polygonal vertical cracks (former horizontal root voids) now filled with pinkish gray (7.5YR-7/2) silty clay; ~10 to 20 inch (25 to 50 cm) wide irregularly shaped quasi-polygons typically filled with silt; weak fine subangular structure; slightly hard, firm to friable, non-sticky and non-plastic; locally silty fragipans; abrupt wavy lower boundary (see *Figure 1*, *Photo 1*, labeled "BURIED PALEOSOL").
Qts Uneven contact K (tonalite bedrock)	7.0 aa 10.0+	3Btb	**Buried Paleosol:** Strong brown (7.5YR-4/6) grading downward into brown to dark brown (7.5YR-4/4) medium to coarse loamy sand with moderate, medium-angular blocky structure decreasing to weak, fine subangular blocky to massive structure near base; slightly hard to hard, firm to friable at depth; sticky to slightly sticky decreasing downward; and slightly plastic and slightly sticky; common, fine to medium reddish brown (5YR-4/4) clay films lining ped faces and bridging mineral grains. The paleosol is a strongly developed argillic horizon grading downward into weathered, slightly stratified, grussified quartz diorite (tonalite) (see *Fig. 1*, *Photo 2* labeled "BURIED PALEOSOL") with isolated, granitic clast "ghosts" to ~4-in diameter decreasing in frequency and size with depth. The 3Btb horizon represents ~100–300 ka of weathering.

1 Section measured and described by RJS ad WJE on January 12, 2013.

2 Exposures are in approximately 13'+ (4 m) deep arroyo cuts (Figure 1, Photos 1 and 2).

3 The upper relict paleosol (2Btb) is strongly developed, and though at or near the modern ground surface, formed mainly under climatic conditions of the past and hence is a "relict paleosol."

4 The buried siliceous duripan (2Bqmb) is locally 1 to 2 inches (2.5 to 5 cm) thick. This silcrete impedes and perches seasonal gravitational water. Modern and paleo-root systems are now filled with fragipan-like younger silts translocated down from the modern surface through and along ped faces within the overlying strong angular blocky structure of the relict paleosol (Figure 1, Photo 2).

5 Our understanding and interpretation of Bush and others (2012, p. 64–66) stratigraphic section.

Photo 1 Photo of site 1.

Photo 2 Photo of site 2.

Photo 3 Photo of site 3.

Photo 4 Photo of site 4.

Photo 5 Photo of site 5.

SELECTED REFERENCES (*Annotated*)

Abbott, P. L., 1999, The rise and fall of San Diego, 150 million years of history recorded in sedimentary rocks: Sunbelt Publications, San Diego, California, 231 p.

Atwater, T., 1970, Implications of plate tectonics for the Cenozoic tectonic evolution of Western North America: Geological Society of America Bulletin, v. 81, p. 3513–3536.

Bahlburg, H. S. M., and Weiss, R., 2010, Comment on "sedimentary features of tsunami backwash deposits in a shallow marine Miocene setting, Mejillones Peninsula, northern Chile" by G. Cantalamessa and C. Di Celma [Sedimentary Geology 178 (2005) 259–273] (2021), p. 77–80].

Bourgeois, J., 2009, Geologic effects and records of tsunamis: *in* Robinson, A. R., and Bernard, N., eds., The sea, tsunamis, v. 15, Harvard University Press, Cambridge, MA, USA, p. 53–91.

Bridge, J. S., 2008, Discussion of articles in "Sedimentary features of tsunami deposits," Sedimentary Geology, v. 211, p. 94.

Brigham-Grette, J., Melles, M., Minyuk, P., Andreev, A., Tarasov, P., DeConto, R., Koenig, S, Nowaczyk, N., Wennrich, V., Rosen, P., Haltia-Hovi, E., Cook, T., Gebhardt, T., Meyer-Jacob, C., Snyder, J., and Herzschuh, U., 2013, Pliocene warmth, polar amplification, and stepped Pleistocene cooling recorded in northeast arctic Russia: Science Express http://www.sciencemag.org/content/early/recent/9 May 2013/Page 1/10.1126/science.1233137

Busch, L. L., Swanson, B. J., Hernandez, J. L., and Olson, B. P. E., 2012, 2nd printing, Preliminary search for Eltanin impact tsunami deposits in southern California–A possible early Pleistocene chronozone: *in* Gomes, C., ed., Waiting for tsunami, coastal hazards of northern San Diego County, California: San Diego Association of Geologists annual field trip guidebook, September 8 and 9, 2012, distributed by Sunbelt Publications, El Cajon, CA., p. 57–81. (Includes five location maps, five color photographs, and six black and white photographs.)

Cantalamessa, G., and Di Celma, C., 2005, Sedimentary features of tsunami backwash deposits in a shallow marine Miocene setting, Mejillones Peninsula, northern Chile: Sedimentary Geology, v. 178, p. 259–273.

Chamberlin, T. C., 1890, The method of multiple working hypotheses: Science (old series), v. 15, p. 92–96.

Dawson, A. G., and Shi, S., 2000, Tsunami deposits: Pure and applied geophysics, v. 157, p. 875–897.

Donnelly, M., 1934, Geology and mineral deposits of the Julian District, San Diego County, California: California Journal of Mines and Geology, v. 30, n. 4, p. 331–370. (Includes description of regional physiography and the old erosion surface (peneplain) of Gastil, 1961)

Ellis, A. J., and Lee, C. H., 1919, Geology and ground waters of the western part of San Diego County, California: United States Geological Survey Water-Supply Paper 446, 321 p., approximate map scale (Pl. 3) 1" = 4 miles.

Gastil, R. G., 1961,The elevated erosion surfaces: *in* Thomas, B. E., ed., Guidebook for field trips, Cordilleran section of The Geological Society of America and associated societies, 57th annual meeting, San Diego State College, fall, 1961, p. 1–3a.

Gastil, R. G. and Bushee, J., 1961, Geology and geomorphology of eastern San Diego County [California], field trip n. 1, road log: *in* Thomas, B. E., ed., Guidebook for field trips, Cordilleran section of The Geological Society of America and associated societies, 57th annual meeting, San Diego State College, fall, 1961, p. 8–22. (Reference to Lakeview Mountain tonalite underlying sites 1, 2, and 3. Sites 4a, and 4b, are underlain by Bonsall tonalite.).

Gersonde, R., Kyte, F. T., Frederichs, T., Bleil, U., and Kuhn, G., 2003, New data on the late Pliocene Eltanin impact into the deep southern Ocean: Third international conference on large meteorite impacts, August 5–7, 2003, Nördlingen, Germany, 4094.pdf, 2 p.

Goff, J. R., and Dominey-Howes, D., 2010, Does the Eltanin asteroid tsunami provide an alternative explanation for the Australian mega-tsunami hypothesis?: Natural hazards and earth system sciences, v. 10, p. 713–715.

Goff, J., Chagué-Goff, C., Archer, M., Dominey-Howes, D., and Turney, C., 2012, The Eltanin asteroid impact—Possible South Pacific palaeomegatsunami footprint and potential implications for the Pliocene-Pleistocene transition: Journal of quaternary science, v. 27, n. 7, p. 660–670.

Google Earth images, 2017 (world wide web).

Haq, B. U., Hardenbol, J., and Vail, P. R., 1987, Chronology of fluctuating sea levels since Triassic: Science, v. 235, p. 1156–1167.

Hernandez, J. L., Todd, V. R., Busch, L. L., and Tan, S. S., 2007, Geologic map of the San Pasqual 7.5′ quadrangle, San Diego County, California: A digital database: California Geological Survey (1:24,000). ftp://ftp.consrv.ca.gov/pub/dmg/rgmp/Prelim_geo_pdf/SanPasqual_prelim.pdf

Hertlin, L. G., and Grant, U. S., IV, 1944, The geology and paleontology of the marine Pliocene of San Diego, California, part I, geology: San Diego Society of Natural History memoir, v. 2, p. 1–72.

Hinds, N. E. A., 1952, Evolution of the California Landscape: California Division of Mines Bulletin 158, 240 p. [Accompanied by: Jenkins, O. P., 1938, Geomorphic Map of California: California Division of Mines, Bulletin 158, Pl. 2, map scale 1:2,000,842, contour interval 2,000′.]

Hubbard Scientific, Inc., 1979, Santa Ana Sheet, 1° x 2° plastic raised relief topographic map n. NI-11–8, scale 1:250,000, contour interval 200′ (Hubbard Scientific, Inc., Chippewa Falls, WI.).

Jahns, R. H., 1954, Geology of the Peninsular Range Province, Southern California and Baja California, *in* Jahns, R. H., ed., Geology of Southern California: California Division of Mines, Bulletin 170, chapter 2, contribution 3, p.29–52. (Includes: Bulletin 170, chapter 2, plate 3, generalized geologic map of the Peninsular Range Province, Southern California.)

Jenkins, O. P., 1938, Geomorphic map of California: California Division of Mines, Bulletin 158, Pl. 2, map scale 1:2,000,842, contour interval 2,000′.

Kominz, M. A., 2001, Sea level variations over geologic time: Academic Press, p. 2605–2613. doi: 10.1006/rwos.2001.2055

Kyte, F. T, Gersonde, R., and Kuhn, G., 2009, Eltanin: ground truth for a kilometer-sized deep ocean impact, American Geophysical Union fall meeting, December 14–18, 2009, San Francisco, CA, abstract NH31D-04.

Le Roux, J. P., and Vargas, G., 2005, Hydraulic behavior of tsunami backflows: insights from their modern and ancient deposits: Environmental Geology, v. 49, p. 65–75.

Melles, M., Brigham-Grette, J., Minyuk, P., Nowaczyk, N. R., Wennrich, V., DeConto, R. M., Anderson, P. M., Andreev, A. A., Coletti, A., Cook, T. M., Haltia-Hovi, E., Kukkonen, M., Lozhkin, A. V., Rosen, P., Tarasov, P., Vogel, H., and Wagner, B., 2012, 2.8 million years of Arctic climate change from Lake El'gygytgyn, northeast Russia: Science 337, 315–320.

Merriam, R., 1946, Igneous and metamorphic rocks of the southwestern part of the Ramona [30′] Quadrangle, San Diego County, California: Geological Society of America Bulletin, v. 57, n. 3, p. 223–260, Pl. 1, map scale 1:160,000 . (Reference to Lakeview tonalite underlying sites 1, 2, and 3. Sites 4 and 5 are underlain by Bonsall tonalite.)

Merriam, R. H., 1954, [Geologic map of the] Ramona-Julian area [CA], *in* Jahns, R. H., ed., Geology of southern California: California Division of Mines Bulletin 170, geologic guide n. 5, map n. 10, p. 43, map scale 1:203,000.

Miller, W. J., 1935, Geomorphology of the southern Peninsular Range of California: Geological Society of America Bulletin, v. 46, n. 10, p. 1535–1562.

Miller, Kenneth G., 2009, Sea level change, last 250 million years: Encyclopedia of Paleoclimatology and Ancient Environments–part of the series, Encyclopedia of Earth Sciences Series, p. 879–887.

Morton, R. A., Gelfenbaum, G., and Jaffe, B. E., 2007, Physical criteria for distinguishing sandy tsunami and storm deposits using modern examples: Sedimentary Geology, v. 200, p. 184–207.

Paine, M., 2017, Tsunami from asteroid/comet impacts: Australian Spaceguard Survey, 10 p., on line at: http://users.tpg.com.au/users/tps-seti/spacegd7.html#tsunamiimpact

Pinter, N., and Ishman, S. E., 2008, Impacts, mega-tsunami, and other extraordinary claims: Geological Society of America Today, v. 18, n. 1, p. 37–38.

Rogers, T. H., compiler, 1966, Geologic map of California, Santa Ana sheet: California Division of Mines and Geology, map sheet, map scale 1:250,000, contour interval 200′. (Reference to Green Valley tonalite at all four locations.)

Sagan, C., 1995, Wonder and skepticism: Skeptical Enquirer, v. 19, p. 24–30.

Sauer, C., 1929, Land forms in the Peninsular Range of California as developed about Warner's Hot Springs and Mesa Grande: University of California publications in geography, v. 3, n. 4, p. 199–290.

Sears, F.W., 1950–58, Principles of physics series, mechanics, heat, and sound: Addison–Wesley Publishing Company, Inc., Reading, Massachusetts, USA, 564p.

Shlemon, R. J., 1978, Late Quaternary evolution of the Camp Pendleton-San Onofre State Beach coastal area, northwestern San Diego County, California: for Southern California Edison Company (Rosemead, CA) and San Diego Gas & Electric Company (San Diego, CA) , Consultant's Technical Report, 128 p.

Shlemon, R. J., 1987, The Cristianitos fault and Quaternary geology, San Onofre State Beach, California: *in* Hill, M. L., ed., Geological Society of America Centennial Field Guide-Cordilleran Section, p. 171–174.

Simpson, D. R., 1965, Geology of the central part of the Ramona Pegmatite District, San Diego County, California: California Division of Mines and Geology special report 86, p. 3–23, map scale 1" = 1 mile.

Thomas, B. E., ed., 1961, Field trip guidebook, San Diego County–1961: Prepared by the geology department at San Diego State College [University], March, 1961, for the 57th annual meeting, Cordilleran section, Geological Society of America and associated societies, 71p.

Todd, V. R., Busch, L. L., Foster, B. D., Hernandez, J. L., and Tan, S. S., 2006, Geologic map of the Ramona 7.5′ quadrangle, San Diego County, California: A digital database: California Geological Survey (1:24,000). ftp://ftp.consrv.ca.gov/pub/dmg/rgmp/Prelim_geo_pdf/Ramona-prelim.pdf

Todd, V. R., Busch, L. L., Hernandez, J. L., Foster, B. D., and Tan, S. S., 2012, Preliminary geologic map of the Ramona 7.5′ quadrangle, San Diego County, California: California Geological Survey, digital database version 2.0, map scale 1:24,000, contour interval 20′. http://www.conservation.ca.gov/cgs/rghm/rgm/preliminary_geologic_maps.htm (Reference to Japatul Valley tonalite [biotite-hornblende and hornblende-biotite tonalite] at sites 1 and 2. Reference to Las Blancas tonalite [hypersthene-biotite tonalite] at site 3. Sites 1 and 2 are found in erosional windows surrounded by late to middle Pleistocene colluvial deposits. Sites 4 and 5, are underlain by tonalite of Alpine, exposed in a window of Pleistocene colluvial deposits.)

United States Department of Agriculture, 1973, Soil survey, San Diego area, California: U. S. Dept. Agriculture, 118 p., map sheet n. 37, Ramona Quadrangle, map scale 1:24,000.

United States Geological Survey, 1947, Ramona, California, 15′ topographic quadrangle map, map scale 1:62,500, contour interval 100′.

United States Geological Survey, 1955, Ramona, California, 7–1/2′ topographic quadrangle map, scale 1:24,000, contour interval 20′.

United States Geological Survey, 1958–78, San Diego, California, 1° x 2° topographic map sheet, N1 11–11, scale 1:250,000, contour interval 200′.

United States Geological Survey, 1959–69, Santa Ana, California, 1° x 2° topographic map sheet, N1 11–8, scale 1:250,000, contour interval 200′.

Walawender, M. J., 2000, The Peninsular Ranges — A geological guide to San Diego's back country: Kendall/Hunt Publishing Company, USA, 114p.

Wardlaw, B. R., and Quinn, T. M., 1991, The record of Pliocene sea level change at Enewetak atoll: Quaternary Science Reviews, v. 10, p. 247–258.

Ward, S. N., and Asphaug, E., 2002, Impact tsunami — Eltanin: Elsevier Science Ltd., Pergamon, Deep Sea Research Part II, 49, p. 1073–1079.

Weber, H. F. Jr., 1963, Geology and mineral resources of San Diego County, California: California Division of Mines and Geology, county report 3, 309 p., Pl. 1, map scale 1" = 2 miles. (Reference to quartz diorite (tonalite) at all four locations.)

Weiss, R., Lynett, P., and Wünnemann, K., 2015, The Eltanin impact and its tsunami along the coast of South America–insights for potential deposits: Earth and Planetary Science Letters, Elsevier B. V., p. 175–181.

www-1: 2017, Eltanin impact: https://en.wikipedia.org/wiki/Eltanin_impact

www-2: 2017, SMS tsunami warning: http://www.sms-tsunami-warning.com/

A Geological Hike into the Santa Ysabel West Open Space Preserve, October 2017

MIKE THACKER

THE AREA

The Santa Ysabel West Open Space Preserve is a park within the County of San Diego that is managed by the County Parks and Recreation Department. It is located along Hwy 78, about 1.4 miles west of the intersection of Hwy 78 and Hwy 79 in Santa Ysabel. This morning's hike (~2.6 miles round trip) will take us about half way into the park, through open meadows dotted with rock outcrops of various types and populated with impressive groves of Engelmann Oak. Of special interest in this area is the huge Ramona Plutonic Complex (*Figure 1*), a structure consisting of 12 Cretaceous plutons arranged in a nearly concentric pattern covering approximately 250 sq. km (Todd et al., 2014).

Google Earth image, a portion of the Santa Ysabel West Open Space Preserve

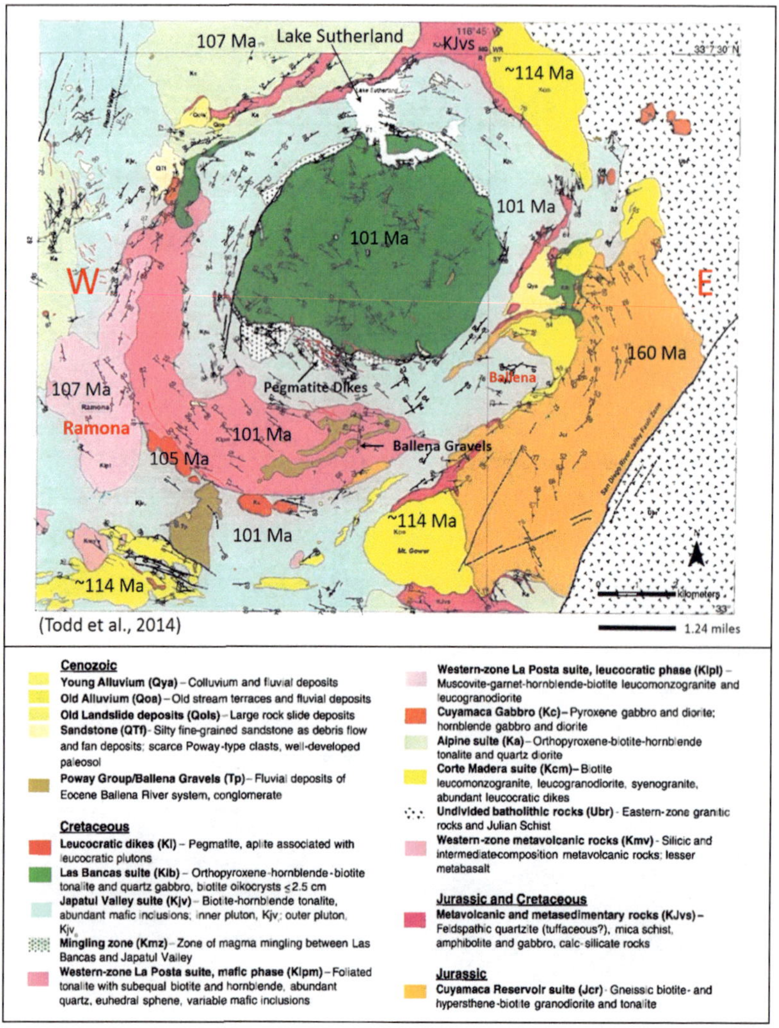

Figure 1. Ramona Plutonic Complex (RPC). Numbers shown are rock emplacements given in the text of Todd et al., 2014. The KJvs unit within the RPC has not yet been dated. However, KJvs samples from southern San Diego County yielded U-Pb dates of 165 Ma, 150 Ma and 110 Ma. The Kcm plutons within the RPC have also not been dated. Two samples of Kcm from El Cajon Mountain immediately south of the RPC produced ages of 114 Ma and 115.6 Ma.

If you arrived here by traveling Hwy 78 from Ramona, you entered the complex shortly after leaving Ramona, drove through its center (at the turnoff to Lake Sutherland), only leaving it approximately 2000 feet west of the trailhead. In *Figure 1*, this pluton is identified as Las Bancas tonalite (Klb), although it is sometimes labeled Green Valley tonalite on other geological maps.

Also of geological significance is the Psuedogravity gradient (PSGR) line (Langenheim and Jachens, 2003) that passes through the Ramona Plutonic Complex (*Figure 2*). This line represents the suture zone at depth, between magnetic western batholithic rocks and essentially nonmagnetic eastern batholithic rocks. It is derived mathematically in a manner that integrates or smooths upper-to middle-crustal intrusive contacts (Todd et. al, 2014). The suture zone also represents the western boundary of Middle to Late Jurassic metagranites (Shaw et al., 2014), such as the Cuyamaca Reservoir suite shown

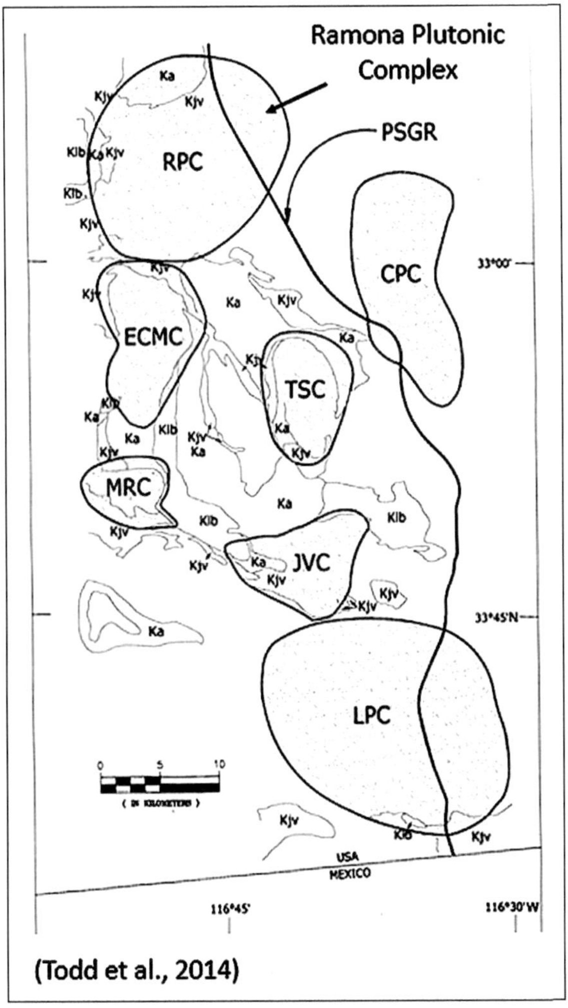

Figure 2. The location of several plutonic complexes within San Diego County, including the Ramona Plutonic Complex. The pseudogravity gradient line (PSGR) passes through the RPC.

in *Figure 1*. Within the RPC, metavolcanic and metasedimentary rocks (unit KJvs) have been interpreted as being deposited in a backarc basin between Jurassic arcs. This basin closed during latest Jurassic(?)–Early Cretaceous time, as a western Jurassic volcanic arc was accreted to the continental margin (Shaw et. al, 2014). Note that unit KJvs in the northeast portion of the RPC (*Figure 1*) is located very near the path of the PSGR shown in *Figure 2*.

To sum up, what we have in this area is a batholithic suture between western and eastern batholithic rocks and the western boundary of Jurassic metagranites in San Diego County where a backarc basin closed as a Jurassic arc accreted to the continental margin. This suture zone was then intruded into and deformed by the Cretaceous plutons of the Ramona Plutonic Complex, in essence, stitching the whole assembly together like a giant geologic needle and thread.

THE HIKE

We'll start the hike in an area of mixed granitic and metamorphic rock with prominent outcrops of tonalite (unit KJem), then we'll move into an area of significant Triassic-Jurassic Julian Schist (JTrm). Of note are nearby zones of gabbroic rock (unit Kc). Also, as we pass through groves of Engelmann Oak, be aware that they are one of the rarest trees in the world, everywhere except San Diego. Due to the impacts of climate change over millennia, the formation of deserts and recent pressure from human development, habitat for this species has been reduced such that now over ninety percent of Engelmann Oak stands are in San Diego County (Pavlik et al., 1991)

STOP 1: Parking lot. Granitic rocks of unit KJem (tonalite) with mafic inclusions and quartz-feldspar dikes.

Here we'll see rock that is characteristic of local granitic outcrops. The silicic dikes were formed by injection of water-rich quartz and feldspar rich magmatic fluids from the intrusive body.

Nevertheless, here the East Mesa tonalite pluton (KJem) presents a classic interaction between tonalite and mafic rock showing contact between two co-evolution-ized magmas and the incipient stages of mingling. The basaltic dike shown above is typical for this pluton. Similar basaltic dikes intruded all of the other plutonic suites, even gabbro plutons! Because their chemistry appears to differ from that of the large gabbro plutons, there may be bodies of slightly compositionally more evolved basaltic dike magma at depth. The East Mesa plutons occur only in the central part of the batholith, where one

Photo 1. Granitic rock with erosionally resistant dikes.

Photo 2. Granitic rock criss-crossed with dikes.

Photo 3. Interface between mafic and more felsic rock. I'm afraid the rock pictured here is a couple of hundred yards north of the trail (from the trailhead). Because we're a large group, we won't be able to pay it a visit. However, county rangers who lead hikes in the preserve often go over to this area, offering the opportunity to take a look at this rock.

pluton is of Middle Jurassic age (same as Jcr and Jhc) and a second is Early Cretaceous. This West Santa Ysabel Open Space Preserve pluton is part of one of the largest KJem plutons. Many have dike like shapes with NW trend. Some are mylonitic. Also, they may have been emplaced during a period of extension in the intrusive history of the batholith.

STOP 2: Outcrop with grinding feature (slick)—the relevance of geology to the Kumeyaay.

In addition to being geologically interesting and a nice place to hike, this area is also an archaeological site. Native Americans (Kumeyaay) lived here and left clues to their presence if you know what to look for.

Along the trail you may find flakes of quartz or aphanitic volcanic or metavolcanic rock–flakes produced by the process of stone tool making. Because force travels more efficiently and more predictably through rock with very fine texture, the smaller the crystals are in rock, the better it is for use in tool making. Aphanitic volcanic and metavolcanic rock (very fine grain to microcrystalline) produces a sharp and durable edge, but can be difficult to work with (in my opinion). Cryptocrystalline rock, such as chert or chalcedony, is easier to work with, but these materials are much less common in San Diego than the volcanics or quartz. Obsidian (volcanic glass) is by far the easiest to work with, but this material is brittle and tool edges dull more quickly than with other materials. Also, unfortunately for the Kumeyaay in San Diego, the closest obsidian source is Obsidian Butte on the southeastern shore of the Salton Sea. As an indication of the desirability of good quality obsidian, samples sourced from as far away as Oregon have been found in the Southern California archaeological record.

Another sign of Kumeyaay presence is the milling features we see on bedrock. Different types of features were used for different purposes. Holes in the rock (mortars) were used with pestles to grind acorns and process other foods and materials. Very shallow or flat smooth surfaces, called slicks, were used with hand held grinding stones (manos) in a scraping motion to grind seeds, clays, fibers and other materials. On the bedrock outcropping at this stop we see a slick. You can identify a slick (as you might guess) by its very smooth texture relative to surrounding stone. Generally, milling features will be found on hard bedrock. If very weathered or gritty stone were to be used for a milling surface, grit would get into the food–not "good eats". And rocks with a view seem to be more likely to have milling features on them than rocks without a view (seriously).

STOP 3: Metasedimentary rock exposure by the trail.

We've now entered an area with significant exposures of metasedimentary rock (JTrm) - some poking up at odd angles and some lying flat. Here, just off the trail, is a flat exposure of metasedimentary rock with a variety of textures, inclusions and folds. What are some suggestions about what this rock is telling us about its past?

STOP 4: View of Santa Ysabel Creek and eastern edge of the Ramona Plutonic Complex; and approximate contact between the Julian Schist and a small gabbroic body to the north.

From here we can look north into the valley below and see Santa Ysabel Creek as it flows from its headwaters on Volcan Mountain toward Lake Sutherland. This creek is the beginning of the San Dieguito River that flows into Lake Hodges, eventually to empty at Dog Beach in Del Mar. From here we can also see the north-eastern edge of the Ramona Plutonic Complex. Three miles to the northwest, a light colored rock outcrop can be seen juxtaposed against a brown colored outcrop. This is the interface between the Corte Madera suite (Kcm) and Jurassic and Cretaceous metavolcanic and metasedimentary rock (KJvs) of the RPC. There you can see remnants of the backarc basin closure, suture formation and subsequent intrusion by Cretaceous plutons discussed above.

On the hike back, try to identify RPC plutons in the distance to the south and west. Things tend to look different when viewed from the perspective of the hike back.

REFERENCES SITED

Langenheim, V.E., Jachens, R.C., 2003, Crustal structure of the Peninsular Ranges batholith from magnetic data: Implications for Gulf of California rifting: Geophysical Research Letters, v. 30, no. 11, p. 51–1 to 51–4.

Pavlik, Bruce M., Muick, Pamela C., Johnson, Sharon G., Popper, Marjorie, 1991. Oaks of California, 3rd printing, 1995. Cachuma Press. Inc.

Shaw, Stirling E., Todd, Victoria R., Kimbrough, David L., Pearson, Norman J., 2014, A west-to-east geological transect across the Peninsular Ranges batholith, San Diego County, California: Zircon 176 Hf/177Hf evidence for the mixing of crustal- and mantle–derived magmas, and comparisons with the Sierra Nevada batholith. Geological Society of America, Memoir 211, 499–536.

Todd, Victoria R., Hernandez, Janis L., Busch, Lawrence L., 2014, The Zoned Ramona Plutonic Complex: An Early Cretaceous mid- to upper-crustal intrusive sequence, Peninsular Ranges batholith, southern California. Geological Society of America, Memoir 211, 583–608.

"Julian, Where Goest Thou?"
A Brief History of How Past and Present Plate Motions are Estimated, Ending with the GPS Velocity of our Campsite in Julian, California

MONTE MARSHALL
Professor of Geology and Geophysics, Emeritus
San Diego State University

INTRODUCTION

In my previous articles on plate tectonics, I discussed how the theory was developed and applied to the western US. But I never discussed how relative or absolute plate motions are measured/estimated in a quantitative way, either for the geologic past or at present. So, in this article I will begin by tracing the use of key ideas, like Euler poles and Mercator projections, in the first efforts to quantify plate motions. The formation of the Atlantic Ocean will be used as a good example of plate divergent motions around Euler poles. Next, I will describe the evolution of various models of plate motions, and discuss what geologic data were used to locate the Euler poles and calculate the angular velocities that quantify the relative motions between plates. Then I will show data on current and geologic displacement rates in the Pacific-North American plate boundary zone of Southern California. Finally, I will show how much modern GPS data tell us about current plate motions. And, as a fun, finishing touch, I will focus on the current motion of an appropriate point on this field trip—Julian, California! :>)

THE USE OF EULER POLES AND MERCATOR PROJECTIONS

The first critical step in being able to quantify (write an equation for) plate motions was made by one of the great Cambridge geophysics professors, Sir Edward ("Teddy") Bullard. An early supporter of continental drift, he realized in 1964 that if he wanted to close the Atlantic Ocean and quantitatively show how well the coastlines of the circum-Atlantic continents fit, he would be dealing with the motions of curved surfaces on the face of the earth, not just flat surfaces. Fortunately, he knew that one of the world's most famous mathematicians, Swiss-born Leonhard Euler, had derived equations in the 1700s to handle such geometric problems. Euler's (pronounced "oiler") theorem states that every displacement from one position to another on the surface of the earth can be regarded as a rotation around the pole of a suitably chosen axis passing through the center of the earth. He showed that the rate of separation, r, of two originally adjacent points on a sphere is given by the simple equation: $r = \Omega R \sin\Phi$, where Ω is the angular velocity (usually expressed in radians or degrees/unit time) about a pole somewhere on the sphere, R is the radius of the sphere, and Φ is the angular distance between the pole of rotation and the two points (*Fig. 1*). The same equation holds true for the relative motion of two points on either side of a spreading center, or the absolute motion of an individual point on a single plate. The effect of the sine function is important since it shows that the relative or absolute motion is zero at the Euler pole and increases to a maximum at 90 degrees from the pole. For Euler's theorem to work well in plate tectonics, the plate motion must be horizontal and the plates must behave perfectly rigidly. The lack of rigidity caused some problems in using Euler's theorem in these early years, and, as we shall see, still does.

Several years later one of Bullard's grad students, Dan McKenzie, along with one of his classmates, Robert Parker, showed that Euler poles could be used to describe the relative motions of the earth's newly discovered plates (McKenzie and Parker, 1967). Parker, a brilliant mathematician and geophysicist, who had just taken a position at the Scripps Institute of Oceanography (SIO), in La Jolla, CA, also showed that if the Euler pole of the rotation between two plates is used as the pole of a Mercator projection, any transform faults would plot as straight, horizontal lines on the map.

The final step was made by Princeton geophysicist, Jason Morgan, who had studied the Pacific Ocean fracture zones. When they were shown to be transform faults, i.e., major strike-slip faults, he realized that the fracture zones must lie on small circles around their Euler pole and that the intersection point of a set of great circles perpendicular to the fracture zones was a very

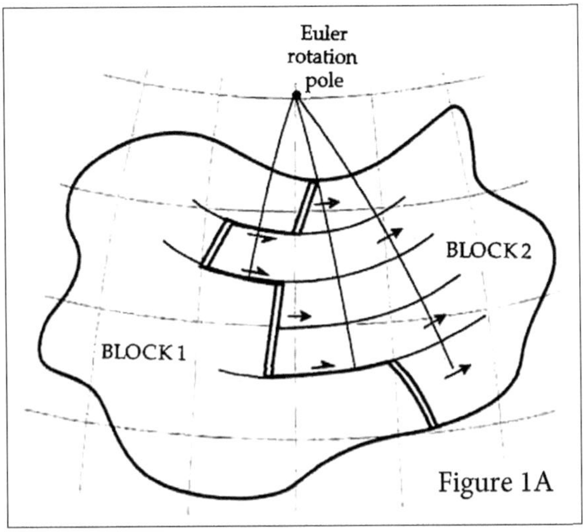

FIGURE 1A. This figure illustrates how the displacement of a rigid plate on the surface of a sphere is described by the rotation of the plate around an Euler pole. Note that the intersection point of great circles drawn perpendicular to the small circles (transform faults) that offset the boundary (double lines/spreading center) is the location of the Euler pole (after Morgan, 1968).

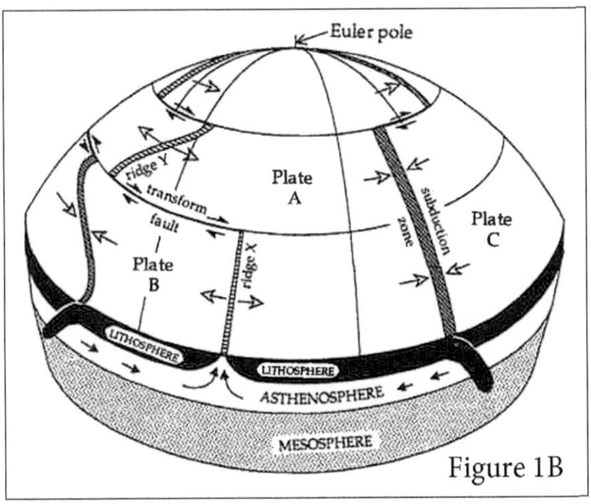

FIGURE 1B. Schematic diagram showing the three types of plate margins, or the three ways that the lithospheric plates can interact—divergent/sea floor spreading ridge, convergent/subduction zone, and transform or strike-slip fault. The figure is drawn relative to the Euler pole of relative motion between plates A and B (Lowrie, 1997).

useful way of finding the location of their Euler pole (Morgan, 1968) (*Fig. 1A*). By this time the linear marine magnetic anomalies had been correlated with the geomagnetic reversal time scale, so seafloor spreading rates were becoming known—giving angular rotation rates for some of the Euler poles. Many of the personal and scientific details in the preceding account are from an excellent book on the early development of the theory of plate tectonics written by my PhD advisor, Allan Cox (Cox, 1973).

As mentioned above, Robert Parker introduced the use of Mercator projections in plate tectonics. Mercator projections, invented by the 16th century Flemish cartographer Gerardus (probably known to his friends as "Jerry"!) Mercator, are one of the most common kinds of maps. Before computers, the only way to accurately represent things on the earth's surface was with a globe. True directions, distances, areas, or shapes on a sphere are distorted when put on a flat surface. From the time of the Greeks, cartographers projected earth's features onto cylinders and cones, and then cut and flattened the projections into planes. Mercator chose to project an earth globe onto a cylinder that was wrapped around it, tangent to the equator, and whose axis was parallel to the earth's. This is the conventional Mercator map widely used in world maps. Since the pole of the projection is the North Pole the great circles of longitude are straight, N-S lines and small circles of latitude are straight, E-W lines. The distortion, which is inherent in all projections, in Mercator maps is that they grossly exaggerate the size of areas in earth's polar regions. This results from two artifacts of the projection. First, the great circles of longitude which converge toward the poles on the earth are parallel lines on the map. Secondly, the small circles of latitude, even if they have the same angular increment, e.g., circles on the earth showing every ten degrees of latitude, become more and more widely spaced horizontal lines on a Mercator map toward higher latitudes. This polar distortion is so great that on a Mercator world map, Greenland has the size of South America! Despite this effect, many of the large world maps are Mercator projections. One big advantage of a Mercator projection is that it is the only projection in which an arc on the earth's surface is a straight line on the map—so this projection is universally used for nautical maps. Since oceanic fracture zones/transform faults lie along small circles (See *Figure 1*), Parker realized that when plotted on a Mercator map they would be straight lines. And, more importantly, if the pole of the Mercator projection, instead of being the North pole as on regular maps, was their Euler pole, transform faults would plot as horizontal, or "E-W", straight lines—just like latitude circles do on conventional maps.

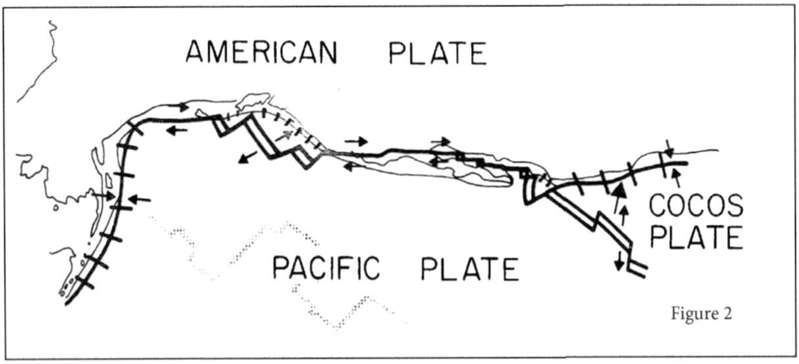

FIGURE 2. A map of the North American-Pacific Plate boundary, with not-yet-subducted remnants of the Farallon Plate—the Cocos Plate off Mexico and the Cascadia Plate off Vancouver Island, B.C., and Washington and Oregon. This map is a Mercator projection, whose pole is the Euler pole for NAP-PP relative motion. The dashed, heavy lines off Alaska and the Aleutian Islands at the left (north) and Mexico at the right (south) are subduction zones. The heavy twin lines are spreading centers and the heavy single lines are transform faults (Atwater, 1970).

Another insightful geophysicist, Tanya Atwater, came to SIO at about the same time as a grad student. In her classic 1970 GSA paper about the ending of subduction of the Farallon plate and the suturing of the Pacific and North American plates, she made a simple Mercator map of the NAP-PP boundary, with the pole of the projection being their Euler pole of relative motion (*Fig. 2*). This map clearly shows what part of the PP-NAP boundary is undergoing subduction (Alaska and the Aleutian Islands), where a remnant of the old Farallon Plate is still subducting (Oregon, Washington, Vancouver Island; and the Cocos plate), and, importantly, that the San Andreas fault, aside from the big bend, is a straight, horizontal line—indicating almost pure strike-slip. The utility of this kind of a map is that any part of the SAF fault, in this case, or any other fault on the map that is part of the plate boundary, must have a component of transpression or transtension if it deviates from horizontal.

THE FORMATION OF THE ATLANTIC OCEAN—CAUSED BY THE DIVERGENCE OF SEVERAL PLATE PAIRS AS THEY ROTATED AROUND THEIR SEPARATE EULER POLES

A good example of plate divergence and how it is controlled by Euler poles is the formation of the Atlantic Ocean during the breakup of Pangea. The first part of this discussion will focus on what plates rifted from what plates and the timing of these separations. The second part will focus on the spreading rates of the mid-Atlantic Ridge (MAR), the role of the oceanic fracture zones/

transform faults, and the Euler poles that governed/still govern their spreading rates. The opening of the Atlantic involved the divergence of three pairs of plates (*Fig. 3*). The northern North Atlantic formed/is forming by the separation of the North American Plate (NAP) from the Eurasian Plate (EUP). The central North Atlantic results from the rifting of the NAP from the African Plate (AFP). The Equatorial and South Atlantic involves the spreading of the South American Plate (SAP) from the AFP. Contrary to expectation, the entire Atlantic Ocean didn't open simultaneously. The earliest plate rifting was between the NAP and the AFP and formed the part of the North Atlantic between Nova Scotia and southern Florida on our side and from northern Morocco to Guinea on the east side (*Fig. 3A*). This rift started in the Jurassic, and, judging by the oldest linear marine magnetic anomalies (LMMA) on both sides, oceanic crust was being formed by 155 Ma. A funny, but geographically true, note is that Jurassic dinosaurs living in the Western Sahara could've walked into Central Park, New York City, to escape the sand! :>)

Determining when two plates began to diverge in the past can be difficult. The conventional model for the formation of a new ocean begins with the stretching and thinning of continental crust accomplished by multiple detachment faults. Closed basins then form and eventually shallow seas invade the sinking crust and evaporites are deposited. More and more basaltic dikes and sills intrude the crust, until they overwhelm the originally felsic rocks along the central axis of rifting/spreading. Even when the new crust becomes entirely basaltic, recognizable LMMAs can't form as long as the basaltic magma intrudes a layer of soft sediments shed from the still nearby continents. For LMMAs to form you need pillow basalt flows emerging from a narrow, central zone of open fissures. This requirement results in magnetic quiet zones, where there are no LMMAs, at the base of continental passive margins. The oldest magnetic anomalies record the beginning of true sea floor spreading/creation possibly tens of millions of years after the adjacent continents had begun rifting apart.

FIGURE 3 (following pages). Map of the mid-Atlantic Ridge in the Atlantic (3A) and Arctic (3B) Oceans. The central pink zone is the oceanic crust created in the last 5 my, and the color-codes for the ages of all the Atlantic oceanic crust are in 3C. The area between the heavy lines in the mid-Atlantic has the oldest crust—the crust formed when Northwest Africa spread from North America. AZTJ, is the Azores triple junction, 15–20FZ is the Fifteen Twenty fracture zone. ENA-EU is the Euler pole for the NAP-EUP spreading center. ENA-AF and ESA-AF are the Euler poles for the spreading centers between the North American and African Plates, and between the South American and African Plates, respectively. Map 3A is a standard Mercator projection, and 3B is a polar, stereographic projection.

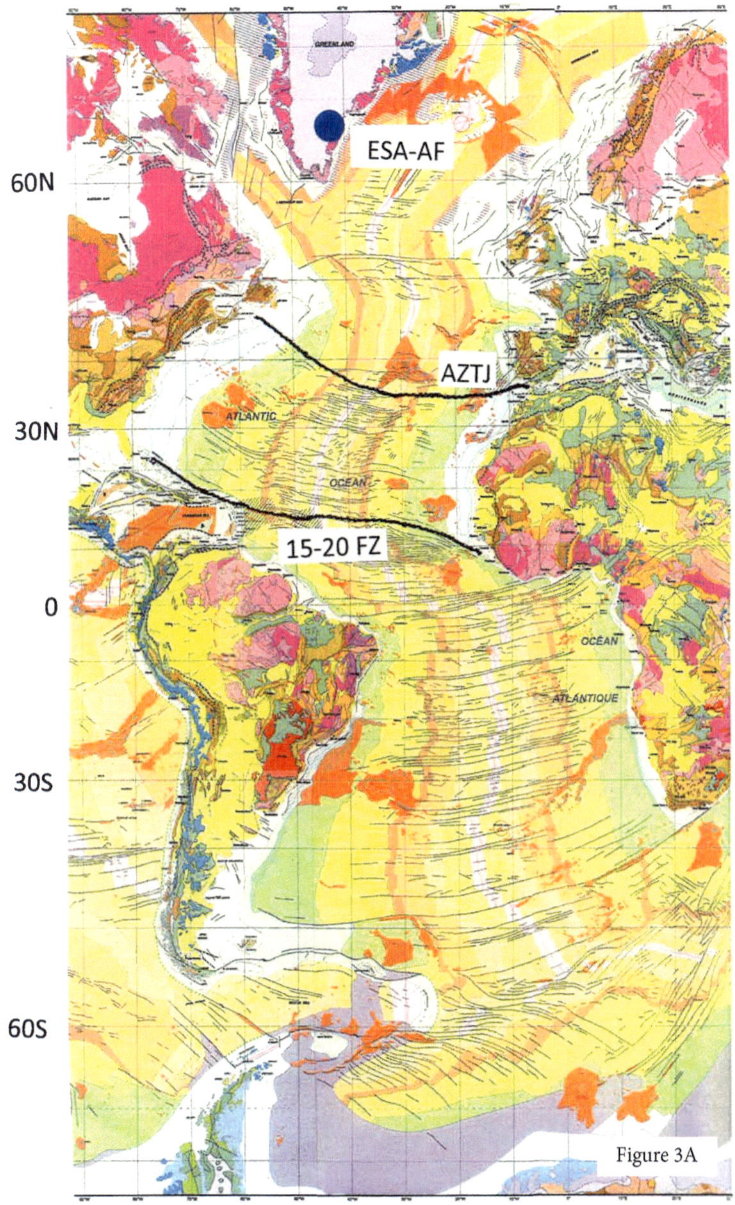
ESA-AF
60N
AZTJ
30N
ATLANTIC
OCEAN
15-20 FZ
0
OCÉAN
ATLANTIQUE
30S
60S

Figure 3A

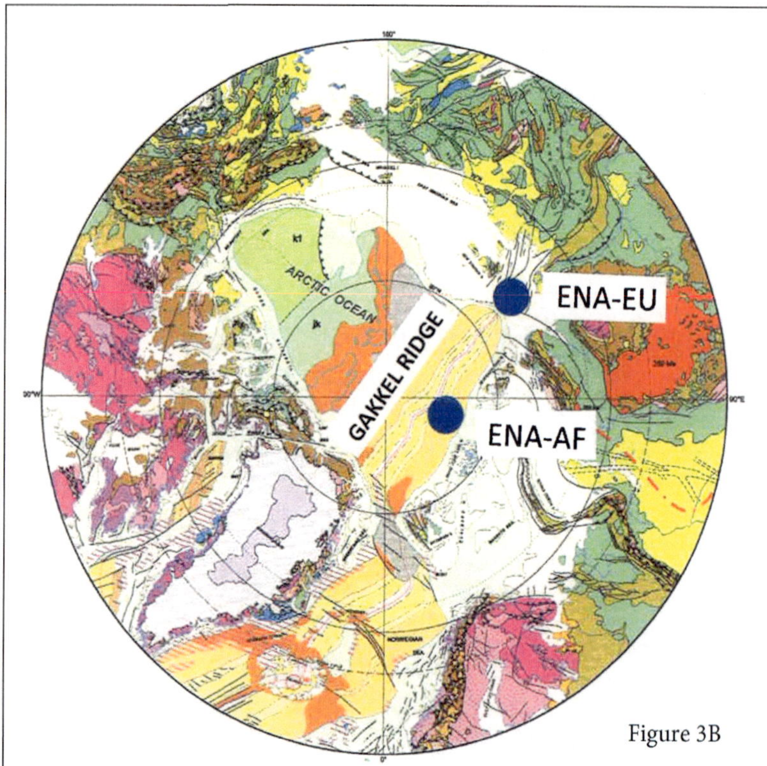

Figure 3B

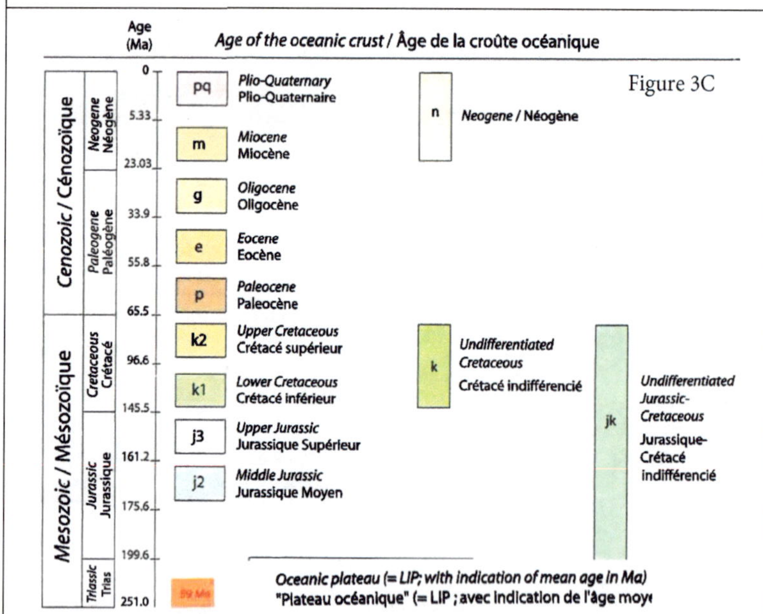

Figure 3C

The next plate pair to rift apart were the South American and African Plates, beginning in the South Atlantic. Based on LMMAs, the crustal extension and thinning between South Africa and Argentina had reached the point where ocean crust was forming by 125 Ma. The spreading center progressed northward, until by 85 Ma, the last link between the SAP and AFP was broken and the equatorial Atlantic formed as the Caribbean separated from Guinea, northwest Africa (*Fig. 3A*). This northward progression is another good example of the fan-shaped boundaries of diverging plate margins caused by increasing separation of the margins with distance from the Euler poles. Oceanic crust was being created in the South Atlantic while there was only continental crustal stretching at the same time to the north.

The last plate pair to begin to separate were the NAP and EUP. Newfoundland Island and northeast Canada had completely separated from Spain and the British Isles by 120 Ma. The last phase of this plate separation and formation of the North Atlantic and Arctic Oceans was complicated by Greenland. Between about 70 Ma and 45 Ma, Greenland rifted from NE Canada to form Baffin Bay. But, beginning at about 55 Ma, the rift between Greenland and Norway became the ultimate plate boundary.

Even though there is only one spreading center, the mid-Atlantic Ridge (MAR), in the center of the Atlantic Ocean, the ocean is the result of the divergence of four large continental plates, two on either side of it, as discussed above. Because the MAR is offset by numerous fracture zones/transform faults and the LMMAs are well mapped, we can construct both the geologic and current Euler poles of the relative motion between the diverging plates. In the following discussion I will use current, GPS-based Euler poles and their angular velocity listed in DeMets, et al. (2010). Starting in the Arctic Ocean, the current Euler pole for the EUP-NAP relative motion is on the Siberian shelf, at the west end of the New Siberian Islands, near the Gakkel Ridge—the northernmost segment of the mid-Atlantic Ridge (*Fig. 3B*). Because it is so close to its Euler pole, the sea floor spreading rate in the Arctic Ocean is very slow. It increases to about 25 mm/year on the MAR at 40 degrees north latitude, at about 65 degrees from the NAP-EUP Euler pole.

At 40N the mid-Atlantic Ridge encounters the Azores triple junction, and whereas its west flank is still the NAP, its east flank becomes the African plate (AFP) all the way to the Antarctic Ocean (*Fig. 3A*). There is a large fracture zone/transform fault at 15N on the MAR, called the Fifteen Twenty F. Z., and south of it the west flank is the SAP (*Fig. 3A*). Therefore, the section of the MAR currently between 40N and 15N is special. This is where the first part of the

Atlantic Ocean opened as northwest Africa rifted away from the east coast of the U.S. and Nova Scotia, as discussed above. This is the NAP-AFP divergent margin, and its current Euler pole is on the Gakkel Ridge, near the North Pole.

In the Equatorial and South Atlantic the spreading rate of any point on the MAR south of 15N is determined by its distance from, and the angular velocity of, the SAP-AFP Euler pole. The current Euler pole is located at 63N, just off the SE coast of Greenland, and has an angular velocity of 0.26 degrees/my, slightly higher than that of the other two poles discussed above. Because of these three Euler poles, the spreading rate of the MAR, and therefore the rate at which the Atlantic Ocean is currently widening, increases from almost zero in the Arctic Ocean, to about 25 mm/y in the mid-North Atlantic, and reaches a maximum of 35 mm/y in the south Equatorial Atlantic, 90 degrees from the SE Greenland Euler pole.

MODELING RELATIVE AND ABSOLUTE PLATE MOTIONS

In the years since the sixties, geodesists and plate tectonicists have discovered more and more plates—from about 12 major plates to more than 60 major, minor, and miniscule plates! With the help of computers and geologic data of varying quality, they have attempted to calculate, or model, the Euler pole and angular velocity of the plate's geologically recent relative and absolute motion. In this section I will summarize the history of the increasingly sophisticated plate motion models that have been devised over the past 40 years. By doing this, the reader will become aware of the kinds of data used in creating these models and their limitations. The modeled motions fall into four categories: relative and absolute, current and geologically recent motions (usually in the last 3 million years). The reasons for a time limit is that the input data becomes less accurate and less complete with time, and Euler poles change over time as the motion of the plates change.

One of the most sophisticated of the early plate motion models (Relative Motion 2, or RM2) was devised by Minster and Jordan (1978). Bernard Minster is currently a geophysics professor at SIO, and Tom Jordan is a geophysics professor and head of the Southern California Earthquake Center (SCEC) at USC. They worked with data on 11 major plates—incorporating 110 seafloor spreading rates, 78 transform fault azimuths, and 142 earthquake slip vectors. They then mathematically inverted these data, and used least squares algorithms to determine a best-fit set of relative motions between the 11 plates. In order to convert their relative motions into absolute vectors they tied the motions to a set of presumably fixed hot spots. Past absolute plate motions are more uncertain than relative motions, because there is no fixed reference point on the earth's surface. The earth's poles of rotation would seem to be good reference points. But the only way a rock can record when it was close to one of the poles is if it's paleomagnetic direction is vertical. The earth's average geomagnetic poles apparently coincided with the geographic poles for most of the Phanerozoic, but there have been times in the past when the earth's rotation axis may have moved relative to the mantle (true polar wander). Absolute, geologic plate motion models have relied on certain hot spots remaining fixed to the mantle for several hundred million years. Although their calculations were done almost 40 years ago, their estimates for the longitude and latitude of the Euler poles and their angular velocities, for the eleven major plates whose relative motions they modeled, are close to the present values.

Subsequent studies have mostly followed Minster and Jordan's (1978) approach, and made refinements to their Euler pole values. One improvement followed an adjustment in the geomagnetic reversal time scale which affected estimates of sea floor spreading rates. The relative motions of smaller plates with their additional geologic constraints have been added to the models making them more complete and have less uncertainty. Some of the conflicting values between different models have been resolved by the addition of new geologic data, i.e., less reliance on earthquake slip vectors when some were found to be ambiguous, or when some motions were found to be erroneous because the plates/parts of plates didn't act rigidly. Modern models have adopted an NNR assumption—that there is no net rotation of the global plate system relative to the mantle. Along with these refinements, model names have changed—with names like NUVEL-1, NUVEL-1A, REVEL, and, more recently, MORVEL. MORVEL is an acronym for mid-ocean ridge velocity, and incorporates 25 plates, and assumes no net rotation (DeMets, et al, 2010).

MEASURING GROUND MOTION IN THE SOUTHERN CALIFORNIA NAP-PP BOUNDARY ZONE

The heart of the NAP-PP boundary is the San Andreas fault system and it is especially complicated in Southern California, because of its many fault strands. One of the earliest and most sophisticated studies of the relative motions between these fault strands was done by Saucier and Humphreys (1993). They derived values for the current displacement rate across the fault strands using space geodetic (VLBI, which will be explained in the next section) measurements, and used geologic data for past relative motions. They also used the NUVEL-1, NAP-PP rate of 49 mm/y as a constraint on their total offset rate. *Figure 4A* shows their Quaternary fault offset-based model, and 4B shows their mostly geodetically-based (VLBI) model. The long-term and short-term slip rates for most of the fault strands are similar. The notable exceptions are the Agua Blanca fault zone and its northward extension up to the Alta California off-shore faults and the Eastern California Shear Zone (ECSZ) in the eastern Mojave. In both cases, the modern slip rate is higher than the geologic—they were uncertain whether there has been a real increase in slip rates, or if the difference is the result of limited geodetic and geologic data in these two regions. Given their 12 mm/y of cumulative offset on the faults off-shore of Southern California, they pointed out the increased earthquake hazard to coastal communities. One point in which they were more confident, is an apparent geologically recent extension of the NAP-PP boundary zone to the east of California, for an additional 2 mm/y. *Figure 5* is a newer geologic fault slip rate by Tom Rockwell (2010), which presumably incorporates the fault trench data that he, his students, and other researchers have measured. These geologic slip rates are more consistent with Saucier and Humphrey's VLBI data.

THE ADVENT OF SPACE GEODESY

The last three decades have seen the emergence of space geodesy. Geodesy is the science concerned with the determination of the size and shape of the earth and the precise location of points on its surface. Formerly limited to measurements made entirely by ground-based instruments, now objects in space, especially satellites, have vastly improved this science. Accurately measuring the length of a long line on the earth's surface was very difficult until recently. The first important contributions from space geodesy were made by pairs of radio telescopes that could measure very accurately the time difference in the arrival of plane radio waves from distant quasars. Called VLBI, or very long baseline interferometry, the distance between two points hundreds of miles apart could be measured to a few millimeters. With the advent of satellites, space geodesy took a giant step forward, especially when

FIGURE 4 (opposite). Fault slip rates for the most important faults in Southern California in mm/y. The slip rates in Fig. 4A are from a geologically-based model, and those in 4B are from a largely geodetic (VLBI) model. These maps are Mercator projections whose pole is the NUVEL1 Euler pole of NAP-PP relative motion. The distortion in geographic points is due to the Mercator projection. The fault zones in 3A are: ECSZ, Eastern California Shear Zone; SA, San Andreas; SJ, San Jacinto; IM, Imperial; EL, Elsinore; RC, Rose Canyon; CB, Coronado Bank; NI, Newport-Inglewood; PV, Palos Verdes; SDT, San Diego Trough; AB, Agua Blanca. In Figure 4B, SD and JU are San Diego and Julian, respectively. (Saucier and Humphreys, 1993).

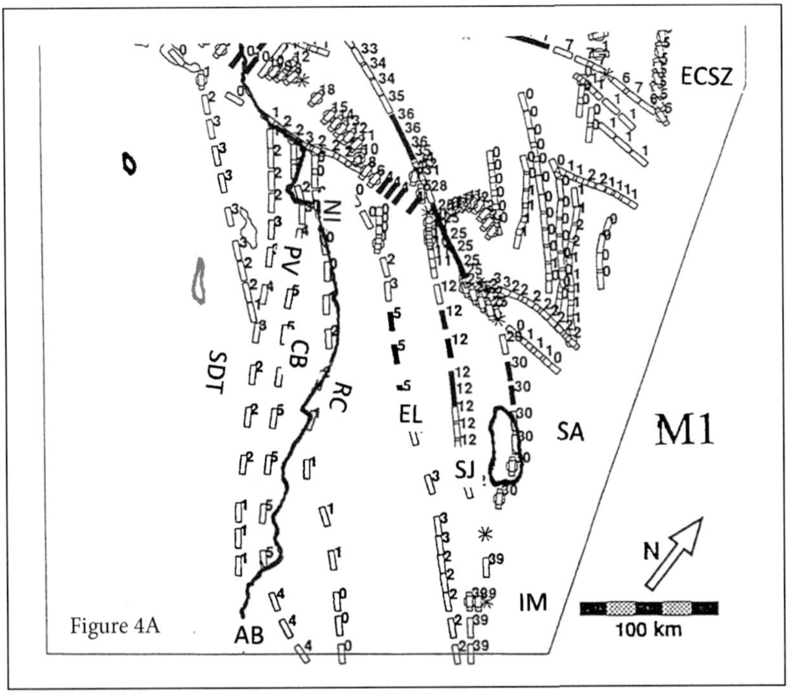

Figure 4A

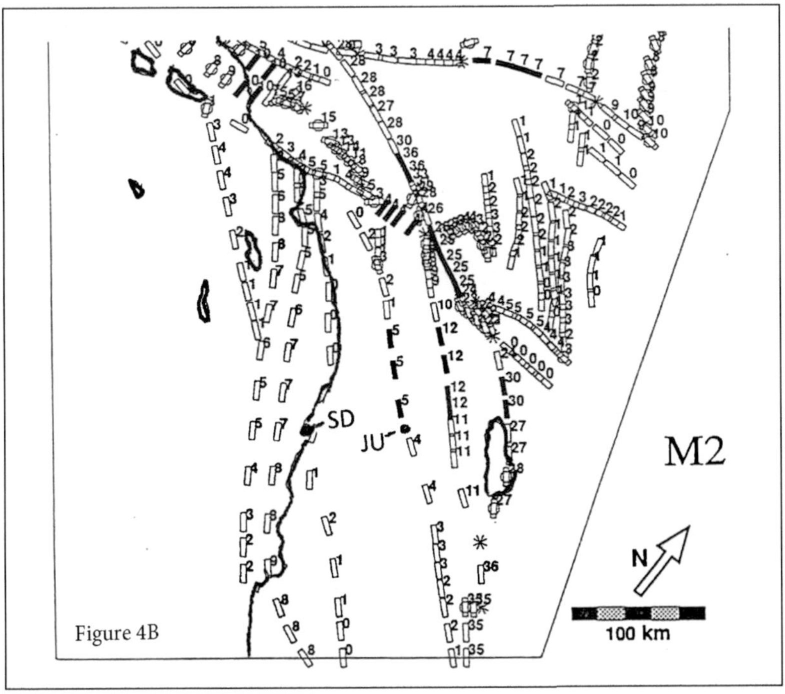

Figure 4B

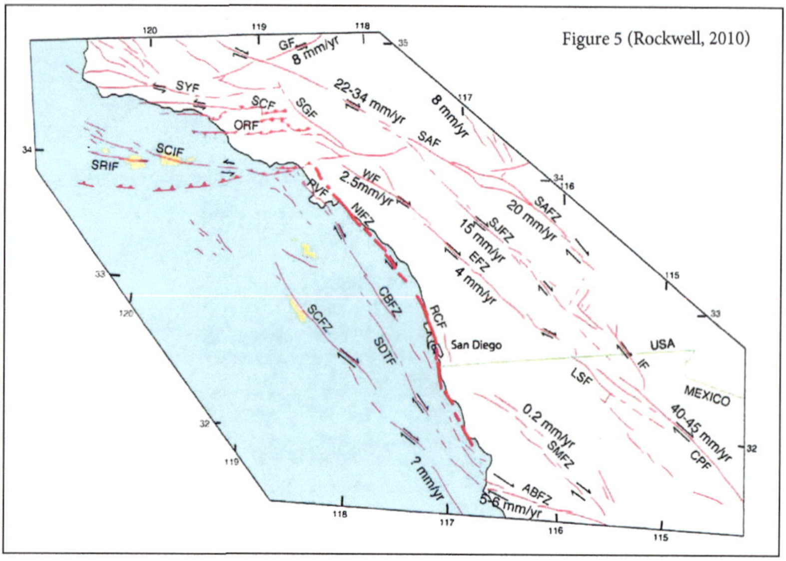

FIGURE 5. Generalized map of major faults in southern California, across which there is about 50 mm/yr of relative motion. The Rose Canyon Fault (RCF) is bolded. Other faults are as follows: SAFZ = San Andreas Fault Zone, SJFZ = San Jacinto Fault Zone, EFZ = Elsinore Fault Zone, LSF = Laguna Salada Fault, CPF = Cerro Prieto Fault, ABFZ = Agua Blanca Fault Zone, SMFZ = San Miguel Fault Zone, IF = Imperial Fault, WF = Whittier Fault, PVF = Palos Verdes Fault, ORF = Oak Ridge Fault, SCF = San Cayetano Fault, SGF = San Gabriel Fault, SRIF = Santa Rosa Island Fault, SCIF = Santa Cruz Island Fault, SYF = Santa Ynez Fault, GF = Garlock Fault, SDTF = San Diego Trough Fault, SCFZ = San Clemente Fault Zone. CBFZ = Coronado Banks Fault Zone

the military installed the Global Positioning System (GPS). Dozens of satellites currently send radio signals to the ground. A ground receiver/antenna then decodes the radio signals into satellite-antenna distances. The data from one satellite tells you only that you are on the surface of a sphere with a known center and radius. With data from three satellites you know that you are at the intersection "point" of the three spheres. This is a three-dimensional version of the method seismologists use to determine the epicenter of earthquakes. For accurate locations a fourth satellite is needed to solve for the time difference between the receiver and satellite clocks. So, a minimum of four satellites need to be in the sky at the same time above a certain location, and the accuracy increases with increasing number of satellites and their azimuthal distribution/coverage. BTW, the instruments used in commercial surveying and geodesy cost tens of thousands of dollars, and are not the ones in your car! With the GPS system, latitude and longitude of points on the earth can be measured with almost millimeter accuracy. Current plate motions can be now be measured more accurately than geological ones. In the beginning of plate modelling, absolute geological plate motions had to be derived from relative motions, by tying them to some hopefully fixed point in the mantle (e.g., certain hot spots). Current relative plate motions, are derived from GPS absolute motions. Locally, SIO is one of the major data centers that collects the world-wide motion data. There are currently some 2000 points on the earth where there are either 'permanent' GPS instruments, where the data are continuously collected in real time, or 'temporary' stations that are reoccupied from time to time. Data must be collected over a span of some years, especially on slowly moving plates, to be meaningful. Location of these stations is important. Motion of the central areas of plates is fairly constant, but friction in the elastic, upper crust of the boundary zones between plates will retard this relative motion and bend the plate edges. When this friction is overcome, the two sides slip past one another, and the bend becomes a straighter line with an offset. Therefore, a station in the boundary zone can show a record of years of too little displacement, followed by a possible one or two-meter advance in seconds. This "S"-shaped bending of the ground, followed by fault slip, is a great example of the "stick-slip" earthquake mechanism discovered in the study of the ground displacement in the 1906 San Francisco earthquake.

WHAT GPS DATA TELL US ABOUT CURRENT PLATE MOTIONS—FROM THE WHOLE WORLD, TO SOUTHERN CALIFORNIA, TO OUR CAMPGROUND IN JULIAN

As mentioned, there are some 2000 scientific GPS stations around the world. You can see the data from most of them on a great website: https:sideshow.jpl.nasa.gov/post/series.html. Most of the following figures are from this website. *Figure 6A* is a beautiful map that shows the current velocities of points in the western hemisphere. The red arrows are the absolute velocity, the paired thin black arrows are the relative velocity (Tarbuck, et al., 2011) The curving set of red vectors across North America is one of the best examples of the rotation of a plate around its Euler pole. Note that, contrary to common belief, the direction of absolute motion changes across the U.S. People on the east coast are moving to the WNW, whereas people in the western US, east of California, are moving to the WSW! As is geometrically obvious, the Euler pole for the NAP is in the SE Pacific, about midway between the Galapagos Islands and the west coast of Ecuador. Note that the vector length/velocity increases to the north, away from the pole. Although not as dramatic, you can see the velocity

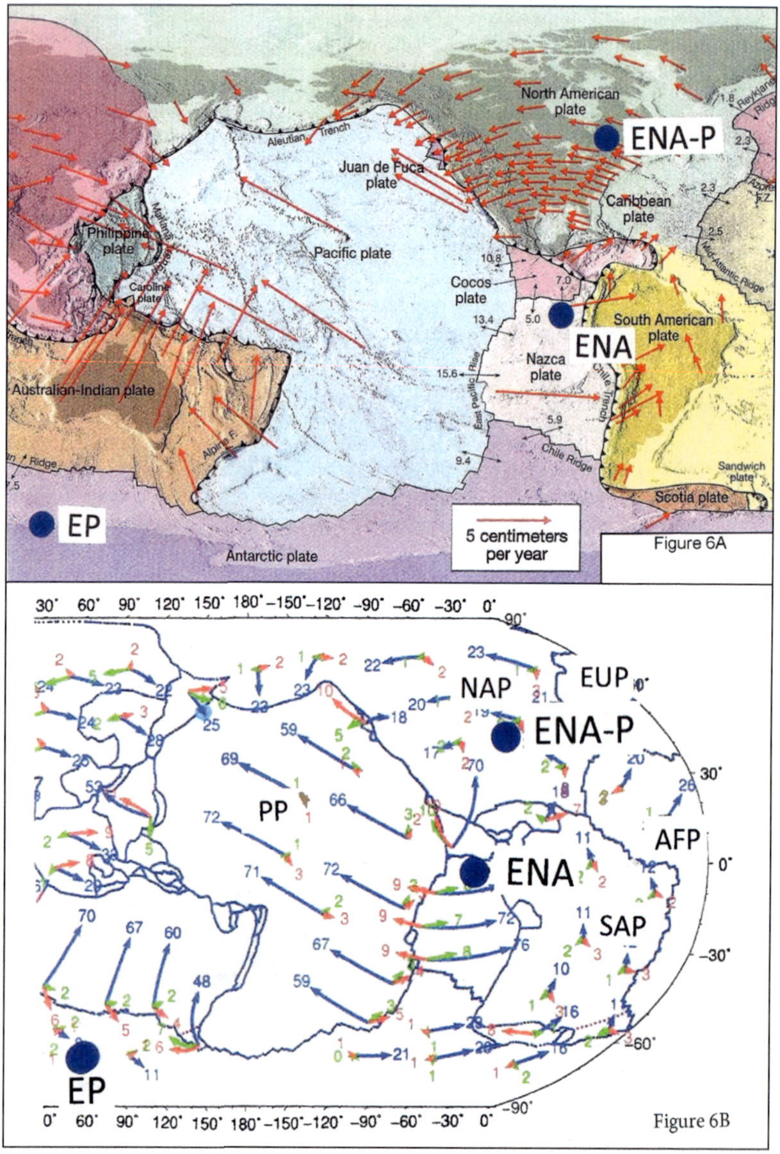

FIGURE 6. 6A is a stunning illustration of the velocity data from GPS stations in the Western Hemisphere and Pacific region. The red arrows show the absolute velocity of the points at the base of the arrows. The black, double-headed arrows show relative motion (Tarbuck, et al., 2011). Here and in Fig. 6B, EP and ENA are the Euler poles for the absolute motion of the Pacific Plate and the North American Plate, respectively. ENA-P is the Euler pole for the relative motion between the two. 6B. For comparison, and because it gives actual velocity numbers, this plate motion map shows the results of modelling of 56 plate motions (Argus, et al., 2011). Unlike 6A, which is based on one or more decades of GPS data, this model is based on sea floor spreading rates over the past 3 million years. The blue arrows and numbers are horizontal velocities, the small red and green arrows and numbers are velocity differences with previous models. PP, Pacific Plate; NAP, North American Plate; SAP, South American Plate; EUP, Eurasian Plate; AFP, African Plate.

of points on the Pacific plate increase and then decrease with increasing distance from its Euler pole, on the coastline of Antarctica. This is another good example of the sine 'effect' on a points distance in degrees from its Euler pole. The third important pole for us is the Euler pole of *relative* motion between the NAP and the PP, since it's the pole for the San Andreas System. It's located in eastern Canada, between the Gulf of St. Lawrence and the southeastern shore of Hudson Bay. For comparison, and because it gives actual velocity values, *Figure 6B* shows absolute plate velocities derived in a modification of the MOVEL model—MOVEL56, which uses geologically recent data from 31 additional micro-plates (Argus, et al., 2011). Here, you can see quantitatively the increase in velocity/arc length as you go north in North America. Note, too, that the velocity of the PP is maximum in the equatorial Pacific (72 mm/y) and then decreases further north, to 59 mm/y in the NE Pacific.

So, finally we come to the center of this year's SDAG fieldtrip—Julian, California! Julian also happens to be fairly near the center of the continental part of the 200-km wide NAP-PP transition zone. *Figure 7* shows how points in central Arizona are moving southwesterly about 2 cm/y. The Pacific seafloor, west of the borderland faults, is moving about 6 cm to the NW. *Figure 8* shows the dividing line between the part of Southern California whose current motion is dominated by that of the NAP, and the part that is dominated by the PP motion. Note in the Imperial Valley that the Calipatria station is moving SW, whereas Westmoreland, just 15 km to the SW is moving to the NW! Evidently,

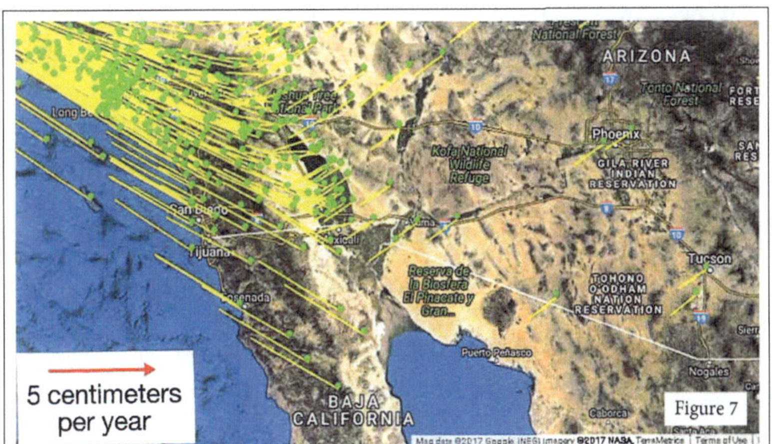

FIGURE 7. GPS velocities measured in SW Arizona and Southern California. Note the SW direction of the North American Plate in SW Arizona and the short distance over which the plate motion changes to the NW velocities of the Pacific Plate (Figures 7, 8, and 9 are from the JPL-NASA website given in the text.).

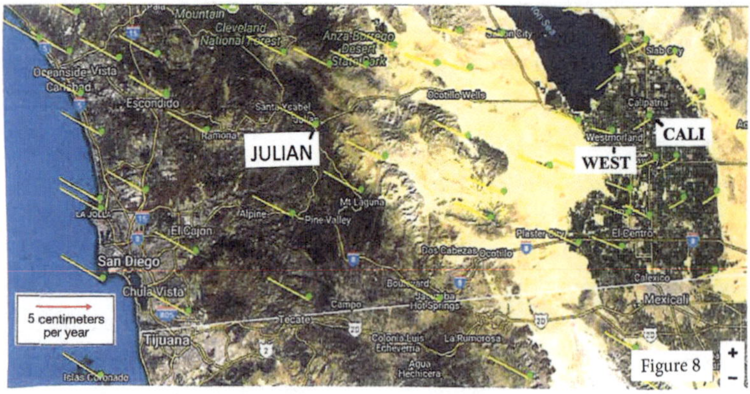

FIGURE 8. GPS velocities in San Diego and Imperial Counties. You can see that the change in plate motions occurs along a line defined by the Imperial Fault and its step-over (rhombochasm) to the northeast shore of the Salton Sea (See fault maps in Figures 4 and 5). Although you might not see any evidence of it at the surface, the towns of El Centro, Imperial, and Westmoreland (WEST) are moving northwesterly with the PP, whereas Brawley and Calipatria (CALI) are more a part of the NAP! At the crest of the Peninsular Ranges, you can see that our home base, Julian, is in the middle of a group of vectors that have the same direction and magnitude.

the northward extension of the Imperial Fault, and the rhombochasm that connects it to the southern end of the SAF on the NE shore of the Salton Sea runs between those two towns! There is something special about this part of the plate boundary zone. Although buried by arrows in *Figure 8*, there is a narrow region along the SAF, north of the Salton Sea, where the GPS station motions point due west. The width of this westward-moving zone increases to the north, and in central California it extends from the southern Sierras across the San Juaquin valley. This is one line of evidence that the NAP-PP boundary zone is broader and more complicated that previously recognized.

You can see Julian, siting at the crest of the Peninsular Ranges, about 60 km NE of San Diego. *Figure 9* shows the GPS data for station P483, located a few kilometers east of Julian. This chart tells us a lot about how the GPS stations operate. It shows 10 years of continuous measurements of changes in latitude (N-S velocity), longitude (E-W velocity) and elevation. Note the huge difference in the scatter of the horizontal components vs the vertical. This is a fundamental property of GPS measurements. The northward component is 16 mm/y, the westward component is 39 mm/y, and there is no significant change in the elevation. Interestingly, however, a sine wave has been fit to the elevation data, and you can see the effect of seasonal changes — the station is about 1 cm higher in the summer, when the rocks are hotter! The vector sum of the two horizontal vectors is 42 mm/y in a N67W direction. Since Julian is in or near the Elsinore fault zone, you might expect these values to possibly be affected by fault-related bending mentioned above. But, you can see that its velocity vector is the same as the stations around it. At a speed of 42 mm/y, if you sleep for 8 hours on this trip, your sleeping bag will be 38 microns, the thickness of a hair, further NW when you get up in the morning! You may better understand all this once the coffee kicks in! :>)

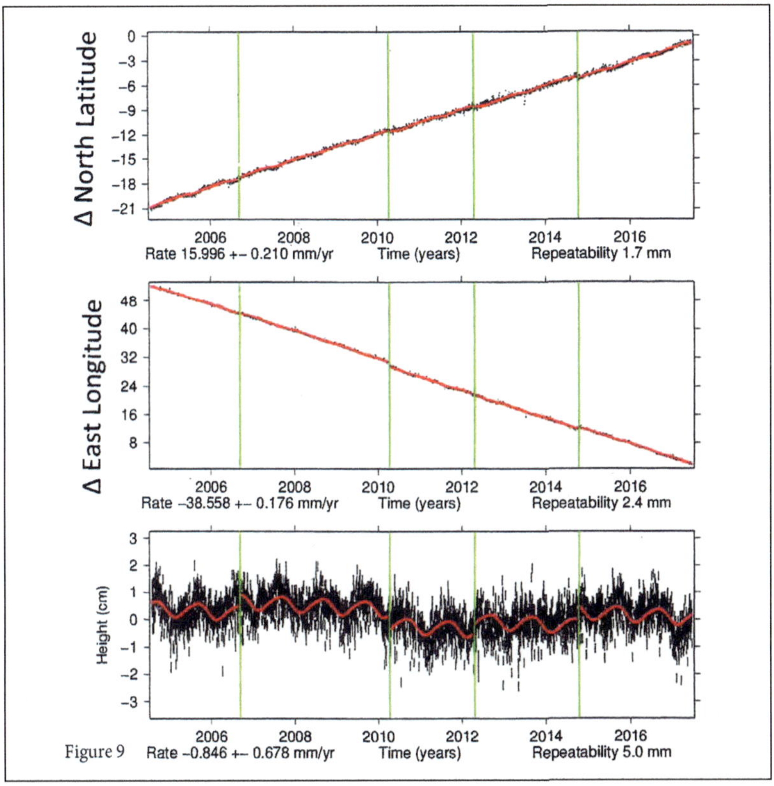

FIGURE 9. Data from GPS station P483, located a few miles E of downtown Julian. This figure shows the three orthogonal components of the station's displacement velocity measured between 2006 and 2016. Δ North Latitude is the northward component, and Δ East Longitude is the westward component. This is a bit confusing until you remember that longitude in many studies is measured only eastward from the Greenwich meridian, 0 to 360 degrees—so a decrease in east longitude amounts to an increase in west longitude, or westward motion! Rounded to the nearest millimeter, you can see that the station's northward velocity is 16 mm/y, its westward velocity is 39 mm/y, and there is no change in its elevation—aside from a seasonal vertical oscillation of one centimeter. The vector sum of the two horizontal components is 42 mm/y in a N67W direction.

ACKNOWLEDGEMENT

The topic for this article was inspired by a recent suggestion of Bill Elliott—a friend for 50 years. We met when we were both seduced into changing our majors (his was mathematics, mine was astrophysics!) to geology by Baylor Brooks at San Diego State College! :>) My interest in things geodetic actually began back in those days, when he got me involved in an on-going departmental project to measure slip on the San Jacinto fault. Each year we students would cart a heavy radio transmitter and receiver up to mountains on either side of the fault and measure the distance between the two points by the delay time in the radio waves. Made by Cubic Corporation of San Diego, it was one of the latest instruments for measuring the distance between two points. I was so impressed by this project that, when the time came to write an essay on the kind of research I was interested in as part of my application for grad school, I wrote about geodesy as applied to fault slip! **And then went on to work on the magnetic properties of sea floor pillow basalt!** I thank Bill for his friendship, his inspiration over the years, and for carefully reviewing this paper—his suggestions have greatly improved it.

REFERENCES

Argus, D. F., R. G. Gordon, and C. DeMets, 2011, Geologically current motion of 56 plates relative to the no-net-rotation reference frame: Geochemistry Geophysics Geosystems, AGU and the Geochemical Society, v. 12, no.11, p. 1–13.

Atwater, T., 1970, Implications of Plate Tectonics for the Cenozoic Tectonic Evolution of Western North America: Geol. Soc. America, Bull., v. 81, p. 3513–3536.

Bouysse, Ph. and coll., 2009, Geological Map of the World at 1:50M (3rd edition), Commission for the Geological Map of the World (CGMW/CCGM), Paris, France.

Cox, A., 1973, Plate Tectonics and Geomagnetic Reversals, W. H. Freeman and Company, San Francisco.

DeMets, C., R. G. Gordon, and D. F. Argus, 2010, Geologically current plate motions: Geophys. J. Int., 181, p. 1–80.

Lowrie, W., 1997, Fundamentals of Geophysics: Cambridge University Press.

McKenzie, D. P. and R. L. Parker, 1967, The North Pacific: an example of tectonics on a sphere: Nature, 216.

Minster, J. B. and T. H. Jordan, 1978, Present-day plate motions: JGR, v.83, no. B11, p. 1276–1280.

Morgan W. J., 1968, Rises, trenches, great faults and crustal blocks: JGR, v. 73, p.1959–1982.

Rockwell, T., 2010, The Rose Canyon Fault Zone in San Diego: Recent Advances in Geotechnical Earthquake Engineering and Soil Dynamics, Fifth International Conference, San Diego, CA, Paper No. 7.06c.

Saucier, F. and E. Humphreys, 1993, Horizontal crustal deformation in Southern California from joint models of geologic and very long baseline interferometry measurements: AGU, Contributions of Space Geodesy to Geodynamics: Crustal Dynamics, p. 139–176.

Tarbuck, E. J., Lutgens, F. K., and Tasa, D., 2011, EARTH, An Introduction to Physical Geology, Tenth Edition, Prentice Hall, Boston.

Julian Schist

Submarine Graveyard of Cratonal North American Mesozoic Fluvial-Systems

JR MORGAN

4671 Lee Avenue, La Mesa, CA 91942 jrmorgan07@cox.net

DAVID L KIMBROUGH

Department of Geological Sciences, San Diego State University, San Diego CA 92182

MARTY GROVE

Department of Geological & Environmental Sciences, Stanford University, Stanford CA 94305–2210.

ABSTRACT

The Julian Schist and related rocks are interpreted as submarine fan turbidites and comprise a major component of the prebatholithic framework of the Peninsular Ranges batholith. Detrital zircon U-Pb ages reported here provide chronostratigraphic and provenance constraints on the Julian Schist. The youngest age peak defined by the U-Pb dates provide an Upper Triassic (228 Ma) maximum depositional age for the Julian Schist protolith. Detrital zircons in the Julian Schist come from three major sources: 1) the Cordilleran arc, 2) the Appalachian-Ouachita orogen, and 3) southwestern North America Yavapai-Mazatzal cratonal basement. These zircons and associated sands were transported from their sources by west flowing transcontinental fluvial and aolian deposits and were ultimately deposited into offshore submarine fans systems by sediment gravity flows. The unequivocal North American provenance of Julian Schist detrital zircon rules out tectonic models that invoke collision of intra-oceanic arc terranes along this segment of the Peninsular Ranges margin.

INTRODUCTION

Detrital zircon U-Pb dating has emerged as a powerful technique in sediment provenance studies, especially with the advent of ion probe and laser ablation analytical methods (Gehrels, 2012). Several factors combine to make zircon U-Pb dating such a valuable method. First of all zircon is a common accessory mineral that crystallizes in a wide variety of igneous and metamorphic. A high diffusion closure temperature of ~900°C for U parent and Pb daughter isotopes make it particularly useful for determining the crystallization age of rocks. Uranium has two major isotopes that independently decay at different rates ($^{238}U \rightarrow ^{206}Pb$ and $^{235}U \rightarrow ^{207}Pb$) and the existence of these two independent geochronometers allows for rigorous evaluation of the closed system requirement for radiometric dating. Finally, zircon is a durable mineral highly resistant to chemical and/or physical alteration so that it is commonly found in sedimentary rocks where it is recycled via weathering and erosion as a 'resistate' heavy mineral useful in provenance studies.

Here we apply detrital zircon U-Pb dating to a study of the Julian Schist in San Diego County (*Figure 1*). The purpose is to: 1) better constrain the depositional age for the Julian Schist sedimentary protolith, and 2) evaluate potential sources for the Julian Schist.

PREVIOUS WORK

Goodyear (1889) is the first to describe the schists in the Julian area. Merrill (1914) uses the term "Julian Group" for the metamorphic rocks in the Julian area, assigns a Paleozoic age to the rocks and correlates them with the Calaveras Formation in the Sierra Nevada. Hudson (1922) investigates the nickeliferous pyrrhotite deposit south of Julian, introduces the term "Julian Schist Series" and assigns a Triassic age to the rocks based on a fossil ammonite.

Donnelly (1934) describes the gold mineralization in the Julian Mining District and introduces the name "Julian Schist" that was subsequently used by several authors (Merriam, 1946, 1958; Larsen, 1948; Everhart, 1951). Germinario (1982, 1993) provides detailed mapping and lithologic descriptions for the Julian Schist and assigns the rocks a Jurassic-Triassic age. Todd (2015) mapped the Julian 7.5-minute Quadrangle assigning the Julian Schist an age of no-younger than Middle Jurassic (160 to 170 Ma).

GEOLOGICAL SETTING

The Julian Schist is one of several prebatholithic formations found in the northern Peninsular Ranges batholith and occurs within both the eastern and

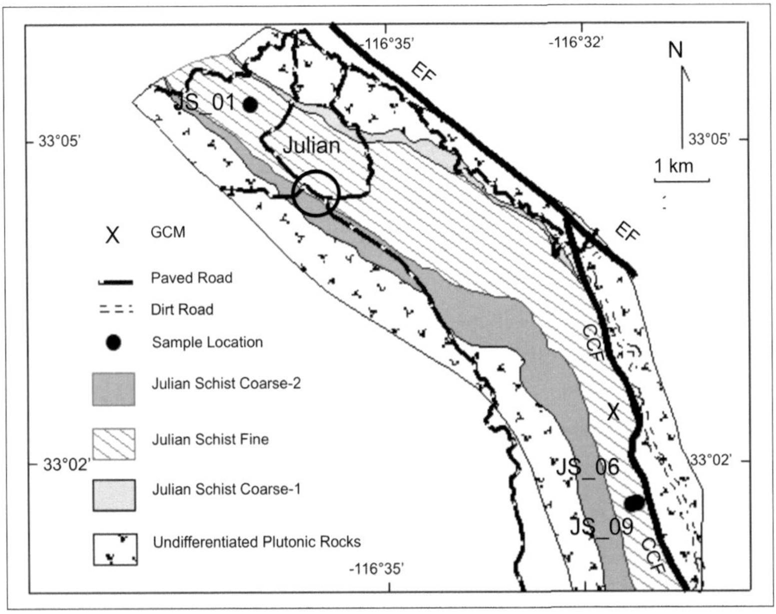

Figure 1. Geologic map of the Julian area modified from Germanario (1982) showing the location of the three Julian Schist detrital zircon U-Pb samples: JS_01, JS_06 and JS_09. JS_01 is quartzite collected from a small hill [Eastwood Hill] northwest of the town of Julian from the same location sampled by Gordon Gastil for detrital zircon U-Pb TIMS dating (Gastil and Girty, 1993). This location is near a shallow prospect-pit, which was probably dug for a mining claim. GPS coordinates are: WGS84 33.091545° -116.613160°. JS_06 and JS_09 are samples of quartzite, which are collected in the bottom of a small creek about 1.55 km south from the Golden Chariot Mine in Chariot Canyon. The two samples were collected about 20 meters from each other. GPS coordinates for the two samples are: JS_06: WGS84: 33.026395° -116.537701° and JS_09: WGS84: 33.026402° -116.537817°. Abbreviations are: CCF-Chariot Canyon Fault; EF-Elsinore Fault; GCM-Golden Chariot Mine.

western zones of the batholith. Other similar rocks include the Bedford Canyon Formation (Larsen, 1948) and the French Valley Formation (Schwarcz,1969), Vallecitos Formation (Reed, 1985) and the Peñasquitos Formation (Kimbrough, et.al. 2014). With the exception of the Peñasquitos Formation, which is predominately dark-gray mudstone with interbedded first-cycle volcanogenic sandstones, the Julian Schist, Bedford Canyon, French Valley and Vallecitos Formations are flysch sediments with thin- to thick-bedded turbidites of sub-arkosic metasandstones interbedded with metapelites. Rarer rock types include chert, amphibolites, gneisses and even-rarer allochthonous marbles (Germanario, 1982). Metamorphism of these rocks ranges in grade from greenschist to upper amphibolite facies.

AGE OF JULIAN SCHIST

The age of the sedimentary protolith of the Julian Schist is uncertain (Todd, 2015). The Julian Schist like most other prebatholithic outcrops in the northern Peninsular Ranges is non-fossiliferous and this lack of fossils has dogged workers for decades. One important exception is a single Triassic(?) ammonite reported by Hudson (1922) in the Julian Schist, which was later lost. Hudson's description of the ammonite find is given below: "A fossil was found by Mr. D. D. Bailey of Julian in a small area of metamorphic rock that lies in the granite about a mile southeast of Banner. This was submitted by Mr. H. L. Huston of San Francisco to Dr. J. P. Smith, who pronounced it *a slender ammonite that is without much doubt Triassic* [oral communication]. The fossil is an imprint of an ammonite on the surface of an angular pebble of dark gray, quartzitic rock. It was found as float. The writer in company with Mr. Bailey visited the locality where this was found. The rocks of the vicinity are non-fissile, quart-two-mica schists, sillimanite gneiss and blue-gray quartzite. Several hours' search was not rewarded by the finding of any fossils, but considerable rock similar in appearance to the matrix of the fossil specimen was seen."

Fossils are also known from multiple outcrops in the Bedford Canyon.

Formation in Ladd Canyon (Siberling, et.al., 1961 and Imlay 1963). Fossils also occur in the Peñasquitos Formation (Fife, et.al., 1967).

The oldest igneous rocks that intrude the Julian Schist are the Cuyamaca Reservoir Gneiss and Harper Creek Gneiss of Jurassic age. Shaw et al. (2003) report zircon U-Pb ages from multiple samples of these units which provide a minimum depositional age for the Julian Schist of 160–170 Ma, since the Julian Schist cannot be younger than the oldest plutons that intrude it.

DETRITAL ZIRCON AGES

Detrital zircon ages are reported here from three samples of Julian Schist (*Figure 1*). All three samples are from the fine-grained facies of Julian Schist as mapped by Germanario (1982, 1993) which is part of the major ~30 km-long northwest-southeast trending screen that underlies the town of Julian and constitutes the type locality of the Julian Schist. The samples were collected from quartz-rich layers of the schist. Separation of zircon from the bulk rock samples and preparation of mounts for analysis was done in the Department of Geological Sciences at San Diego State University. The zircon U-Pb analyses were done at UCLA using the Cameca 1270 ion microprobe. The analytical results are presented in *Table 1* where the age used for each grain is shown in the column labeled "Best Age". For grains less than 1000 Ma years old this is the $^{206}Pb/^{238}U$ age for the zircon and for grains greater that 1000 Ma this is the $^{207}Pb/^{206}Pb$ age which is calculated from the $^{206}Pb/^{238}U$ and $^{207}Pb/^{235}U$ ages.

The detrital zircon U-Pb ages for the Julian Schist samples are shown on a relative probability plot in *Figure 2A*. All of the zircon U-Pb ages from the Julian Schist can be attributed to sources from within the North American craton. A tremendous effort over the past 15 years has gone into characterization of detrital zircon U-Pb age provenance signature of rocks within the region of the Colorado Plateau and more broadly the entire North American Cordillera (e.g. Gehrels et al., 2011; Kimbrough et al., 2014). Guided by this work the Julian Schist age peaks in *Figure 2A* are interpreted to have been derived from three main sources: 1) grains younger than 285 Ma in age are from the Cordilleran magmatic-arc in the USA and/or East Mexico; 2) grains between 399 and 1300 Ma were from Appalachian and Ouachita orogenic sources in southeastern and southern North America and include Grenville age basement (900–1300 Ma), late Neoproterozoic–Cambrian (725–515 Ma), and Paleozoic (510–285 Ma) zircon; and 3) older grains are mostly from the southwestern Cordillera craton from 1.8 to 1.6 Ga Yavapai-Mazatzal basement which is intruded by anorogenic 1.4–1.5 Ga plutons.

The youngest detrital zircon U-Pb ages for the Julian Schist are shown on an expanded scale in *Figure 2B*. The youngest peak of 228 Ma is interpreted as the maximum depositional ages for the Julian Schist.

DISCUSSION

There are no exotic non-North American zircon grains in the Julian Schist amongst the grains analyzed here. All the ages can be accounted for by sources within the North American craton. This result argues against models for the western Peninsular Ranges batholith forming initially as an intra-oceanic arc that was accreted at a later time to western North America (cf. Kimbrough et al., 2014).

The details of the sedimentary pathway(s) by which detrital zircon grains in the Julian Schist were delivered to the continental margin remain a mystery. *Figure 3* modified from Dickinson and Gehrels (2010) that shows the dominant patterns of Mesozoic sediment dispersal across the USA and northern Mexico. As shown here Jura-Triassic North America was characterized by major west to northwest flowing transcontinental river systems originating in the Appalachian-Ouachita orogens on the eastern and southern sides of the continent. The Mesozoic and Cenozoic stratigraphic units on the Colorado Plateau all contain variable proportions of the same three components that make up the Julian Schist detrital zircon population (cf. Kimbrough et al., 2014, 2015). The

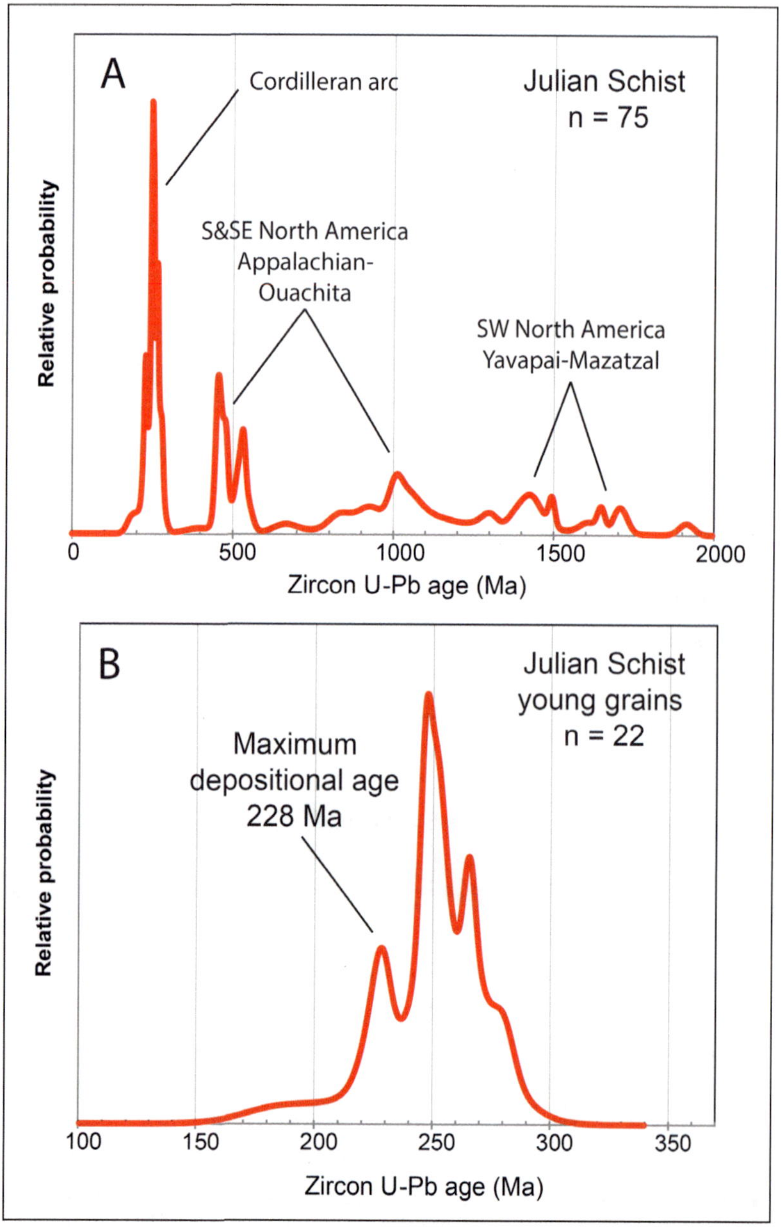

Figure 2. Relative probability distributions of detrital zircon U-Pb ages from the Julian Schist reported here. (A) age distribution for Julian Schist with sources for all the major peaks identified, (B) youngest grains shown on an expanded scale to illustrate the youngest peak (228 Ma) that represents the maximum depositional age for the Julian Schist protolith.

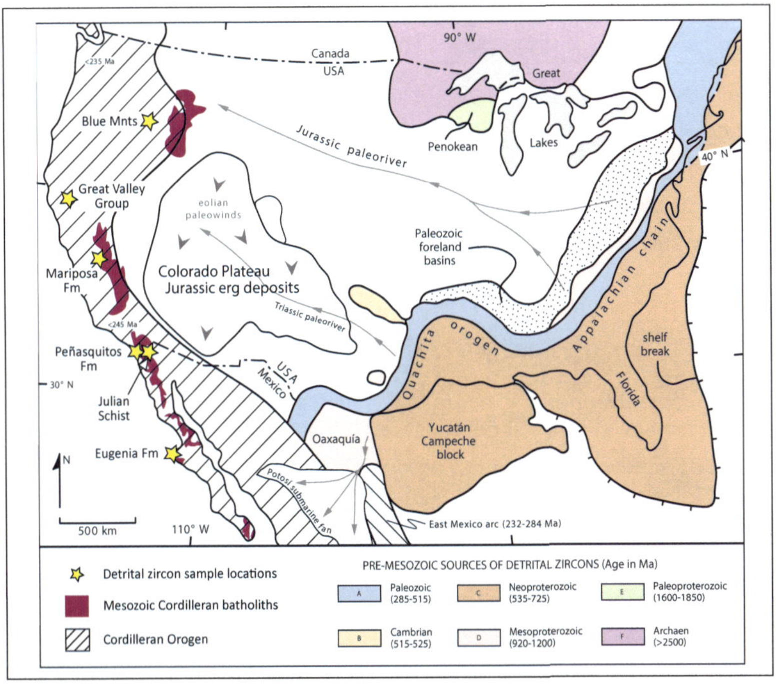

Figure 3. Dominant patterns of Mesozoic sediment dispersal across western North America modified from Dickinson and Gehrels (2010). Detrital zircon sample localities for the Julian Schist and other Jurassic Cordilleran forearc marine deposits are shown here and all share common detrital zircon U-Pb age distributions.

Chinle Formation is part of the northwest flowing fluvial system that traversed the Colorado Plateau and is labeled "Triassic paleoriver" in *Figure 3* (Riggs et al., 1996; 2003; 2016). In north-central Nevada, the Chinle Formation is probably correlative with the marine Osobb Formation. A Triassic shoreline as shown by Riggs may have extended to the southeast toward northwest Sonora. It seems rather problematical that long-shore transport would occur over such a long distance along the Late-Triassic active margin of North America.

A "Jurassic paleoriver" is also shown that traversed the continent to the north of the Colorado Plateau and sands from this river system are thought to have been reworked into the plateau erg deposits. The Late Triassic to Early Jurassic Potosi submarine fan of mainland Mexico also contains a mix of detrital zircon ages similar to the Julian Schist. It is also possible that unidentified through-going rivers cut the Mogollon Highlands and transported sediment to northwestern Sonora. Another possibility is that the dunes from the erg field shown on *Figure 3* transported sand into drainages that leak the zircons into northwestern Sonora into the area where the Julian Schist submarine fan was being deposited.

CONCLUSIONS

Detrital zircon U-Pb ages for the Julian Schist provide a maximum depositional age of ~228 Ma for the sedimentary protolith to the schist based on the youngest grains analyzed. The zircons in the Julian Schist are all attributable to North American sources. There is no evidence from detrital-zircon grains in the Julian Schist that the Peninsular Ranges are part of an exotic far-traveled terrane. If exotic terranes passed by the Peninsular Ranges they left no evidence of their passing in the Julian Schist.

ACKNOWLEDGMENTS

The ion microprobe facility at UCLA is partly supported by a grant from the Instrumentation and Facilities Program, Division of Earth Sciences, National Science Foundation.

REFERENCES

Dickinson, W.R., and Gehrels, G.E., 2009, U-Pb ages of detrital zircons in Jurassic eolian and associated sandstones of the Colorado Plateau: Evidence for transcontinental dispersal and intraregional recycling of sediment: Geological Society of America Bulletin, v. 121, p. 408–433, doi:10.1130/B26406.1.

Dickinson, W.R., and Gehrels, G.E., 2010, Insights into North American paleogeography and paleotectonics from U-Pb ages of detrital zircons in Mesozoic strata of the Colorado Plateau: International Journal of Earth Sciences, v. 99, p. 1247–1265.

Donnelly, M., 1934, Geology and mineral deposits of the Julian District, San Diego County, California: California Journal Mines and Geology, v. 30, p. 331–370.

Everhart, D.L., 1951, Geology of the Cuyamaca Peak Quadrangle, San Diego County, California: California Division of Mines Bulletin, v. 159, p. 51–115.

Fife, D.L., Minch, J.A., and Crampton, P.J., 1967, Late Jurassic age of the Santiago Peak volcanics, California: Geological Society of America Bulletin, v. 78, p. 299–304.

Gehrels, G.E., Blakey, R., Karlstrom, K.E., Timmons, J.M., Dickinson, B., and Pecha, M., 2011, Detrital zircon U-Pb geochronology of Paleozoic strata in the Grand Canyon, Arizona: Lithosphere, v. 3, p. 183–200.

Gehrels, G., 2012, Detrital Zircon U-Pb Geochronology: Current Methods and New Opportunities, in Tectonics of Sedimentary Basins: Recent Advances, C. Busby and A. Azor, editors, Wiley-Blackwell Publishing, p. 47–62.

Germinario, M.P., 1982, The depositional and tectonic environments of the Julian Schist, Julian, California: San Diego, California: San Diego, San Diego State University, [M.S. thesis], 95 p.

Germinario, M., 1993, The early Mesozoic Julian Schist, Julian California, *in* Gastil R.G., and Miller, R.H., eds., The Prebatholithic Stratigraphy of Peninsular California: Boulder, Colorado, Geological Society of America Special Paper 279.

Goodyear, W.A., 1889, 9th Report to the California State Mineralogist, San Diego

County Section, p. 184–305.

Hudson, F.S., 1922, Geology of the Cuyamaca region of California with special reference to the origin of the nickeliferous pyrrhotite: Berkeley, California, University of California Publications Bulletin of the Department of Geological Sciences, v. 13, no. 6, p.175–252.

Imlay, R.L., 1954, Jurassic fossils from southern California: Journal of Paleontology, v. 37, p. 97–107.

Kimbrough, D.L., Abbott, P.L., Balch, D.C., Bartling, S.H., Grove, M., Mahoney, J.B., and Donohue, R.F., 2014, Upper Jurassic Peñasquitos Formation — Forearc basin western wall rock of the Peninsular Ranges batholith, *in* Morton, D.M., and Miller, F.K., eds., Peninsular Ranges Batholith, Baja California and Southern California: Geological Society of America Memoir 211, p. 625–643.

Kimbrough, David L., Grove, Marty, Gehrels, George E., Dorsey, Rebecca J., Howard, Keith A. Lovera, Oscar, Aslan, Andres, House, P. Kyle, Pearthree, Philip A., 2015, Detrital zircon U-Pb provenance of the Colorado River: A 5 m.y. record of incision into cover strata overlying the Colorado Plateau and adjacent regions. Geosphere, Themed Issue: CRevolution 2: Origin and Evolution of the Colorado River System II

Larsen, E.S., 1948, Batholith and associated rocks of Corona, Elsinore, and San Luis Rey Quadrangles, Southern California: Geological Society of America Memoir 29, 182 p.

Merriam. R.H., 1946, Igneous and metamorphic rocks of the southwestern part of the Ramona Quadrangle, San Diego County, California: Geological Society of America Bulletin, v. 57, p. 223–260.

Merriam. R.H., 1958, Geology of the Santa Ysabel Quadrangle, San Diego County, California: California Division of Mines Bulletin, v. 117, p. 7–20.

Merrill, F.A., F.H., 1914, 14th Report of the California State Mineralogist, p. 637–722.

Reed, J.B., 1985, Sedimentology of some Triassic-Middle Jurassic (?) siliciclastic flysch deposits, northwest Baja California Norte, Mexico: San Diego, California: San Diego, San Diego State University, [M.S. thesis], 265 p.

Riggs, N.R., Lehman, T.M., Gehrels, G.E. and Dickinson, W.R., 1996, Detrital zircon link between headwaters and terminus of the Upper Triassic Chinle-Dockum paleoriver system: Science, v. 273, p. 97–100.

Riggs. N.R., Ash, S.R., Barth, A.P., Gehrels, G.E., and Wooden, J.L., Isotopic age of the Black Forest Bed, 2003, Petrified Forest member, Chinle Formation, Arizona: An example of dating a continental sandstone: Boulder, Colorado, Geological Society of America Bulletin, v. 115, n. 11, p. 1315–1323.

Riggs, N. R., Oberling, Z. A., Howell, E. R., Parker, W. G., Barth, A. P., Cecil, M. R., & Martz, J. W., 2016, Sources of volcanic detritus in the basal Chinle Formation, southwestern Laurentia, and implications for the Early Mesozoic magmatic arc. Geosphere, 12(2), 439–463.

Schwarcz, H.P., 1969, Pre-Cretaceous sedimentation and metamorphism in the Winchester area, northern Peninsular Ranges, California: Geological Society of America Special Paper 100, 78 p.

Shaw, S.E., Todd, V.R., and Grove, M., 2003, Jurassic peraluminous gneissic granites in the axial zone of the Peninsular Ranges, southern California, *in* Johnson, S.E.,

Paterson, S.R., Fletcher, J.M., Girty, G.H., Kimbrough, D. L., and Martin-Barajas, A., eds., Tectonic evolution of northwestern Mexico and the southwestern USA: Boulder, Colorado, Geological Society of America Special Paper 374, p. 157–183.

Siberling, N.J., Schoellhamer, J.E., Gray, C.H., and Imlay, R.W., 1961, Upper Jurassic fossils from the Bedford Canyon Formation, southern California: American Association of Petroleum Geologist Bulletin, v. 45, 1746–1765.

Todd, V.R., 2015, Geologic map of the Julian 7.5′ quadrangle, San Diego County, California: U.S. Geological Survey Open-File Report 94–16, 18 p., 1 sheet, scale 1:24,000.

Table 1. Detrital zircon U-Pb ages from the Julian Schist determined using the Cameca 1270 ion microprobe at UCLA.

$^{206}Pb/^{238}U$ Age (Ma)	$^{206}Pb/^{238}U$ ± 1s	$^{206}Pb^*$ (%)	$^{207}Pb/^{235}U$ Age (Ma)	$^{207}Pb/^{235}U$ ± 1s	$^{207}Pb^*$ (%)	U O/ U	U	Th	Th/U	Name	Best age (Ma)	± 1s (Ma)
186	16	99.9	208	24	98.5	8.82	44	30	0.68	JS_09-04-08-01.ais	186	16
217	16	99.7	210	17	94.7	7.96	259	188	0.73	JS_09-04-22-01.ais	217	16
225	5	99.8	224	6	96.1	8.21	544	463	0.85	JS_09-04-03-01.ais	225	5
229	4	98.9	206	11	83.2	9.41	933	460	0.49	JS_01_03-10-01.ais	229	4
232	12	98.9	234	26	85.1	8.49	105	84	0.80	JS_09-04-07-01.ais	232	12
235	10	99.6	234	10	93.2	8.5	318	196	0.62	JS_09-04-20-01.ais	235	10
247	2	99.8	250	6	97	8.2	724	443	0.61	JS_06-02-13-01.ais	247	2
248	12	92.8	289	102	48.1	10.5	192	199	1.04	JS_01_02-01-01.ais	248	12
249	9	99.9	265	10	98.1	8.32	314	107	0.34	JS_09-04-18-01.ais	249	9
250	4	99.7	252	10	95.7	8.42	221	131	0.59	JS_06-02-06-01.ais	250	4
250	5	99.9	253	7	97.8	8.31	339	211	0.62	JS_09-04-06-01.ais	250	5
251	4	99.4	255	15	90.6	9.11	539	240	0.45	JS_01_05-01-01.ais	251	4
252	4	99.7	248	13	95.1	8.15	132	37	0.28	JS_06-02-19-01.ais	252	4
254	10	95.7	266	49	59.1	9.73	335	226	0.67	JS_01_02-11-01.ais	254	10
258	12	99.7	260	15	96.1	8.24	202	109	0.54	JS_09-04-12-01.ais	258	12
261	7	99.9	263	8	98.5	8.38	567	166	0.29	JS_09-04-16-01.ais	261	7
262	12	96.6	268	44	64.2	11	252	29	0.12	JS_01_02-18-01.ais	262	12
266	2	99.8	264	6	96.9	8.29	763	394	0.52	JS_06-02-10-01.ais	266	2
269	5	99.4	261	19	90.2	8.45	115	24	0.21	JS_06-02-22-01.ais	269	5
277	15	91.8	245	93	37.5	10.5	166	75	0.45	JS_01_02-06-01.ais	277	15
277	12	95.1	298	84	56	10.2	165	80	0.48	JS_01_02-08-01.ais	277	12
280	5	98.5	246	10	78.1	8.75	508	216	0.43	JS_01_05-05-01.ais	280	5
399	47	77.0	337	237	14.9	10.3	53	43	0.81	JS_01_02-03-01.ais	399	47
450	7	99.8	453	12	97	8.27	266	189	0.71	JS_06-02-04-01.ais	450	7
455	6	99.9	448	10	99	8.08	191	150	0.78	JS_06-02-16-01.ais	455	6
456	12	98.9	513	37	87.9	10.2	487	387	0.79	JS_01_02-04-01.ais	456	12
462	15	98.4	551	74	83.6	10.2	253	232	0.92	JS_01_03-02-01.ais	462	15
463	9	99.5	450	23	92.5	8.02	69	36	0.52	JS_06-02-15-01.ais	463	9
468	13	99.8	463	14	97.7	8.32	474	365	0.77	JS_09-04-15-01.ais	468	13
471	11	98.8	442	30	83.2	9.27	232	170	0.73	JS_01_05-04-01.ais	471	11
480	6	99.9	473	9	98	8.47	218	186	0.85	JS_06-02-14-01.ais	480	6
490	19	97.3	613	108	76.2	10.1	135	152	1.13	JS_01_02-10-01.ais	490	19

$^{206}Pb/^{238}U$ Age (Ma)	$^{206}Pb/^{238}U$ ± 1s	$^{206}Pb^*$ (%)	$^{207}Pb/^{235}U$ Age (Ma)	$^{207}Pb/^{235}U$ ± 1s	$^{207}Pb^*$ (%)	U 0/ U	U	Th	Th/U	Name	Best age (Ma)	± 1s (Ma)
522	13	99.4	504	13	92.2	8.93	1017	267	0.26	JS_01_03-07-01.ais	522	13
523	17	99.2	556	38	90.5	10.5	360	131	0.36	JS_01_02-05-01.ais	523	17
524	13	98.6	509	34	82.4	9.41	104	44	0.42	JS_01_05-03-01.ais	524	13
530	25	99.0	566	34	88.2	10.2	412	82	0.20	JS_01_02-09-01.ais	530	25
534	7	99.8	529	11	97.6	8.33	200	84	0.42	JS_06-02-20-01.ais	534	7
549	14	98.9	553	31	86.1	9.82	449	98	0.22	JS_01_02-13-01.ais	549	14
664	36	98.1	563	26	75	7.66	632	100	0.16	JS_01_03-04-01.ais	664	36
707	82	98.2	780	97	82.6	9.47	17	15	0.89	JS_09-04-21-01.ais	707	82
825	30	99.0	831	21	89.1	9.25	207	44	0.21	JS_01_05-06-01.ais	825	30
857	55	99.9	857	39	98.5	8.17	177	16	0.09	JS_09-04-14-01.ais	857	55
860	42	97.2	913	78	75.2	9.01	45	13	0.29	JS_01_03-08-01.ais	860	42
912	44	99.1	873	49	89.6	8.01	32	10	0.32	JS_06-02-11-01.ais	912	44
933	32	99.0	933	40	88.8	9.11	106	27	0.26	JS_01_03-09-01.ais	933	32
671	11	99.9	735	12	99.3	8.37	251	27	0.11	JS_06-02-05-01.ais	936	36
947	16	100.0	967	12	99.5	9.03	720	8	0.01	JS_01_05-02-01.ais	1013	18
947	27	98.6	977	25	86.3	8.48	305	76	0.25	JS_01_03-06-01.ais	1045	53
959	55	98.3	997	71	83.9	10.2	146	76	0.52	JS_01_02-12-01.ais	1082	155
964	22	99.7	1074	18	97.4	10.2	671	26	0.04	JS_01_02-14-01.ais	1304	24
968	59	99.9	1185	50	99.1	8.72	84	50	0.59	JS_09-04-10-01.ais	1608	27
996	83	96.9	1083	96	75.1	10.2	36	16	0.45	JS_01_02-17-01.ais	1262	199
1005	10	100	1006	9	99.6	8.16	334	151	0.45	JS_06-02-17-01.ais	1007	24
1020	22	100	1057	17	100	8.17	37	5	0.14	JS_06-02-08-01.ais	1136	45
1032	9	99.9	1021	8	99.4	8.35	445	154	0.35	JS_06-02-07-01.ais	997	19
1044	10	99.9	1047	13	98.8	8.37	436	84	0.19	JS_06-02-18-01.ais	1052	28
1057	53	96.0	872	78	61.6	9.95	179	127	0.71	JS_01_02-15-01.ais	1057	53
1103	109	99.3	1097	83	93.1	8.29	30	31	1.03	JS_09-04-19-01.ais	1084	95
1103	11	100	1218	13	99.7	8.43	271	168	0.62	JS_06-02-09-01.ais	1429	28
1104	12	99.9	1096	13	99.2	8.23	265	40	0.15	JS_06-02-03-01.ais	1081	31
1116	21	100	1090	16	99.6	8.3	225	123	0.55	JS_06-02-18-01@1.a	1039	24
1214	19	99.8	1207	21	97.7	8.43	100	104	1.04	JS_06-02-01-01.ais	1194	52
1261	57	96.0	1262	95	70.8	9.95	288	112	0.39	JS_01_02-01-01@1.a	1266	205
1279	50	99.6	1280	30	96.3	10.1	717	190	0.26	JS_01_03-03-01.ais	1281	37
1293	64	98.6	1417	80	89.7	9.98	137	91	0.67	JS_01_03-01-01.ais	1610	134
1375	44	99.9	1377	29	99.4	8.26	206	129	0.63	JS_09-04-05-01.ais	1379	24
1397	53	99.5	1805	38	97.1	8.96	254	264	1.04	JS_01_03-05-01.ais	2316	15
1408	19	99.9	1414	12	99.3	8.68	201	96	0.48	JS_06-02-21-01.ais	1423	26
1412	14	100	1446	8	99.9	8.45	636	210	0.33	JS_06-02-02-01.ais	1495	10
1422	68	99.7	1430	51	97.3	8.06	26	27	1.04	JS_06-02-12-01.ais	1441	71
1451	100	99.9	1438	66	98.7	8.25	99	56	0.57	JS_09-04-11-01.ais	1420	34
1455	63	100	1535	39	99.8	8.66	787	87	0.11	JS_09-04-17-01.ais	1648	12
1502	109	100	1483	60	100	8.22	84	52	0.62	JS_09-04-17-01.ais	1455	32
1635	51	99.4	1761	32	96.2	10.8	351	236	0.67	JS_01_02-07-01.ais	1914	24
1686	41	99.9	1692	26	99.5	8.24	625	46	0.07	JS_09-04-02-01.ais	1698	17
1693	66	100	1704	37	99.6	8.37	127	81	0.64	JS_09-04-09-01.ais	1718	18

Spessartine Garnet from Ramona, San Diego County, California

BY BRENDAN M. LAURS AND KIMBERLY KNOX

FOR MOST OF THE 20TH CENTURY, THE PRINCIPAL source of fine spessartine was the area around the Little Three mine near the town of Ramona in southern California. Two pegmatites—the Hercules-Spessartite and the Spaulding dikes—yielded most of the estimated 40,000–50,000 carats of facet-grade rough mined from 1956 to 1994. Prized for its bright orangy yellow to yellowish orange color, Ramona spessartine has similar R.I. and S.G. values to those of most other pegmatitic gem spessartines. Typical internal features are wavy two-phase partially healed fractures, negative crystals, needles/tubes, growth patterns, and anomalous birefringence; mineral inclusions are rare. The composition of four samples (ranging from light orangy yellow to yellow-orange) was $Sps_{94.8-88.8}Alm_{4.7-11.0}Gro_{0.5-0.2}$, and the absorption spectra of all samples showed strong peaks for Mn^{2+} and subordinate Fe^{2+} bands.

The Little Three mining area near Ramona, California, has produced a variety of pegmatite gems, most notably an intense orange spessartine garnet (*Figure 1*). The Ramona district lies within the extensive southern California pegmatite belt (*Figure 2*). In the early part of the 20th century, San Diego County was one of the world's major gem-producing areas in terms of variety and quantity of material (Tucker and Reed, 1939). The most sought-after gem was tourmaline, recovered principally from the Mesa Grande and Pala districts. Beryl (aquamarine and morganite), spodumene (kunzite, first discovered at Pala), topaz, and smoky quartz were also important, but spessartine was relatively rare—except in the Ramona pegmatites. The bright orangy yellow to yellowish orange color and brilliance of the Ramona spessartines (*Figure 3*), as well as the fact that this was the premier source of top-quality material for decades, have contributed to a premium on their price that was eclipsed only in the late 1990s by fine spessartine from Namibia and Nigeria (S. Robertson, pers. comm., 2001). Despite the historical importance of Ramona spessartine, very little has been published on its gemological characteristics

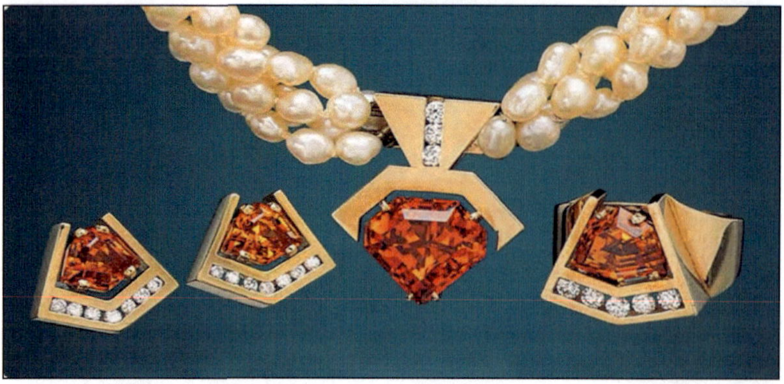

Figure 1. Spessartine from Ramona is prized for its intense orange color. The 10.25 ct spessartine in the pendant was cut from material recovered at the Spessartite mine in 1975. The spessartine in the ring on the right weighs 4.1 ct, and the spessartines in the earrings weigh about 2.5 ct each. Courtesy of Kerith Graeber; photo © Jeff Scovil.

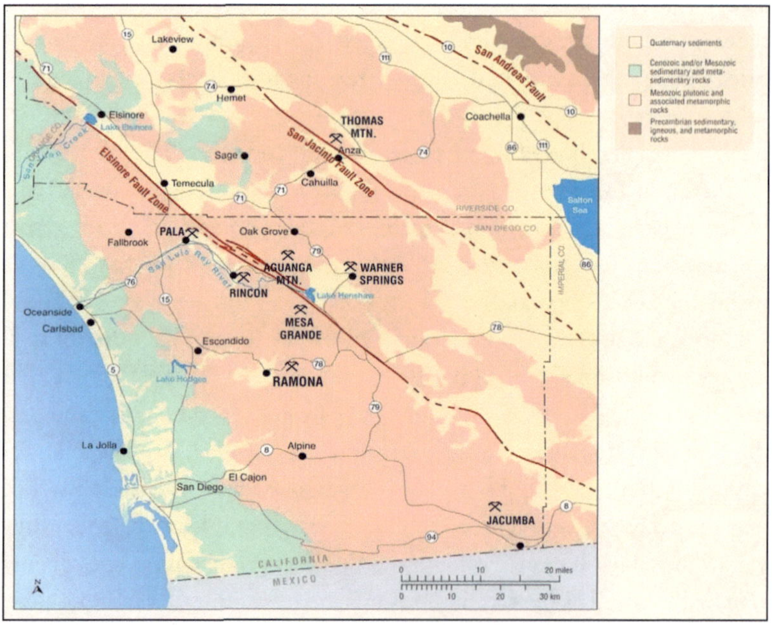

Figure 2. The Ramona district is located in central San Diego County, approximately 50 km (31 miles) northeast of San Diego. Other gem-producing pegmatite districts or localities related to the Peninsular Ranges batholith include Thomas Mountain in Riverside County, and Aguanga Mountain, Jacumba, Mesa Grande, Pala, Rincon, and Warner Springs in San Diego County.

Figure 3. Ramona spessartine (here 1.02–5.15 ct) typically ranges in color from orangy yellow to yellowish orange. Courtesy of Buzz Gray and Bernadine Johnston; photo © Harold and Erica Van Pelt.

(see, e.g., Tisdall, 1961). This article describes the history and production mining methods, geology, and gemology of spessartine from this historic locality. An explanation of the unique conditions necessary for the formation of gem-quality spessartine in granitic pegmatites also is proposed.

HISTORY AND PRODUCTION

Spessartine was not well known as a gemstone until relatively recently. There is no history of the mineral before the 19th century (Rouse, 1986). The first pegmatite mine in the Ramona district, the Little Three, was discovered in May 1903 by H. W. Robb of Escondido, California (Kunz, 1905). It was named for Mr. Robb and his partners, Dan MacIntosh and Charles F. Schnack, who purchased the 40 acre (16.2 hectare) parcel of land from a rancher (Foord et al., 1989; LeMenager, 1989). Although this particular mine has never been an important source of spessartine, its yield of fine specimens of topaz, tourmaline, and other minerals has made it the most important mine in the district. For this reason, Ramona spessartine is often referred to as coming from the "Little Three" mine.

By the end of 1903, several mines and prospects were staked in the Ramona district within an area smaller than 520 hectares (2 square miles). Kunz (1905) provided the best description of the early mines—including the Little Three, A.B.C. (or Daggett), and Hercules (see *Figure 4*). Considerable amounts of topaz, tourmaline, spessartine, and beryl were recovered at the Ramona mines via open cuts and tunnels in the early years (*Figure 5*; see Sterrett, 1908; Tucker, 1925; Tucker and Reed, 1939; Simpson, 1960; Weber, 1963). From 1903 to 1905, Kunz (1905) reported that over 10 kg of spessartine were produced from the district, yielding faceted gems up to 8 ct. By 1909, however, the near-surface deposits at the Little Three mine were considered exhausted (Spaulding, 1977). The collapse of the Chinese tourmaline market in 1912 halted pegmatite mining in southern California (Foord et al., 1989), and there was very little activity for the next four decades.

In 1947, Louis Spaulding and his son, Louis Spaulding, Jr., established several claims in the area (Simpson, 1960). Intensive mining began in 1955 at the Little Three mine, and since then the Spaulding family has been the most important producer of gems and minerals from the Ramona pegmatites. After his father's death in 1973, Spaulding Jr. single-handedly took over work on several of the pegmatites, particularly the Little Three and HerculesSpessartite dikes.

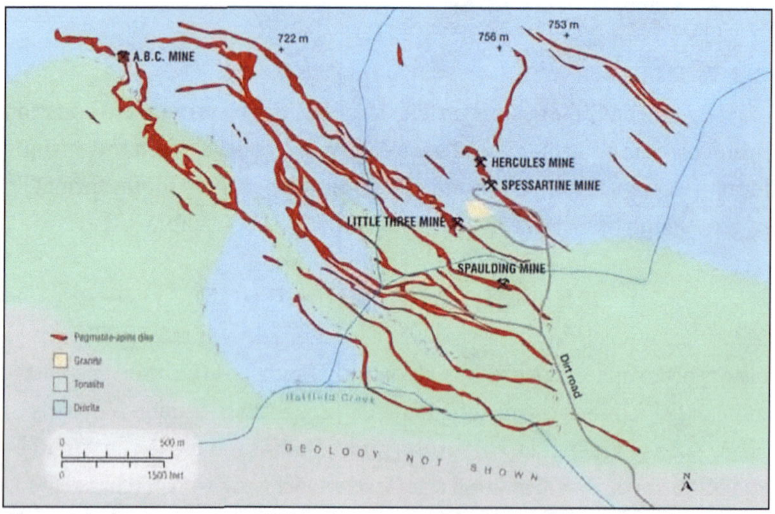

Figure 4. The Ramona district consists of a series of northwest-trending granitic pegmatites that have intruded tonalite and diorite. The locations of the main productive mines are shown: Little Three, Hercules, Spessartite, Spaulding, and A.B.C. Several other claims and prospects are not shown, owing to minor production and/or complexities in their naming and location. Modified from Simpson (1960), Weber (1963), and Foord et al. (1989).

Figure 5. Before the collapse of the Chinese tourmaline market in 1912 halted pegmatite mining in San Diego County, significant gem production was obtained from the Ramona area. The main disturbance is the Little Three mine (left arrow), and the small dumps higher on the hillside are the Hercules mine (right arrow). Photo by D. B. Sterrett, U.S. Geological Survey, circa 1910

Although several mines in the Ramona district have produced spessartine, the best sources were the Hercules-Spessartite and Spaulding mines (again, see *Figure 4*). The Spaulding dike was found in December 1955, and the faceted spessartines produced initially were rather small (0.5–1.5 ct; Sinkankas, 1957). About 3,000 carats of faceting rough were mined over the period 1956–1959 (Sinkankas, 1959; Weber, 1963). From 1964 until shortly before his death, Spaulding Sr. produced 9,000 carats of faceted spessartine from this mine, but only three stones exceeded 6 ct (L. Spaulding Jr., pers. comm., 2001).

The Hercules-Spessartite pegmatite has a longer and more productive history. The Hercules mine was located in August 1903 on the northwest portion of this dike by A. V. Pray, S. G. Ingle, and H. Titus, who worked it for several years (Simpson, 1960). About 15 pounds (6.8 kg) of spessartine were produced during the period 1903–1905 (Kunz, 1905). After mining in the district resumed following the 1912 collapse of the Chinese tourmaline market, owners R. B. Winstead and G. Converse worked the property and even resurveyed it in 1954 (Simpson, 1960), but their production was not recorded. The Hercules was claimed by Elbert McMacken and Herb Walters

Figure 6. The Hercules dike was mined for spessartine by John Sinkankas and his son John William Sinkankas in the mid-1960s. This photo shows Sinkankas Jr. drilling the pegmatite in preparation for blasting. Photo by John Sinkankas, February 1966

in the 1960s and early 1970s; for two years during this period (1965–66), a portion was worked by John Sinkankas and his son (Sinkankas, 1967, 1976; *Figure 6*). Bryant Harris purchased the mine in the late 1970s and worked it with E. "Buzz" Gray until 1984, but with little success (E. Gray, pers. comm., 2001).

The southern portion of the Hercules-Spessartite dike was mined by Spaulding Sr. in 1959; these workings were called the Spessartite mine (Simpson, 1963). This area later became the most important source of spessartine in the district—and produced several stones over 10 ct (see, e.g., *Figure 1*)—when Spaulding Jr. expanded the open cut, beginning in 1975. In August 1987, Spaulding Jr. recovered a 72.5 ct piece of rough that was faceted by Buzz Gray into a 39.63 ct gemstone (*Figure 7*), the largest cut spessartine known to date from Ramona and among the more than 2,300 grams of gem rough produced from 1987 to 1990 (L. Spaulding Jr., pers. comm., 2001). More large stones and mineral specimens (*Figure 8*) were recovered from the Spessartite mine in 1993–1994. Among these was the 27.30 ct gem featured in the *objet d'art* called "The Light of Ramona" (*Figure 9*). Additional small finds of spessartine were made by Spaulding Jr. at various Ramona mines up to 1995, and by another group in 1996–1997 (San Diego Mining Co., 2001), but there has been no significant garnet production since then. Although total production figures are not available, Foord et al.

Figure 7. This magnificent necklace was constructed in 1987 with more than 165 carats of Ramona spessartine (Koivula and Kammerling, 1988). The center stone, at 39.63 ct, is the largest cut spessartine known from Ramona. Courtesy of Michael M. Scott; photo © Harold and Erica Van Pelt

Figure 8. Many large stones and attractive gem crystals of spessartine were recovered from the Spessartite mine during the late 1980s to mid-1990s. The rough and cut (15.93 ct) stones shown here are courtesy of Buzz Gray and Bernadine Johnston; photo © Harold and Erica Van Pelt.

Figure 9. "The Light of Ramona" features a 27.30 ct cushion-cut spessartine accented by 62.23 carats of spessartine, all from Ramona, and 3.81 carats of feldspar. The design replicates the natural isometric crystal habit of garnet (see Figure 8), as conceptualized by Bernadine Johnston and Buzz Gray. The piece is mounted on a covellite pedestal, with more spessartines at the base. It was designed by Kim Knox, and constructed by her together with Zane A. Gillum and Alan F. Hovey of Golden Pacific Arts, San Diego, California. Photos © Harold and Erica Van Pelt.

(1989) estimated that about 40,000 carats of facet-grade spessartine have been recovered from the Ramona district. (Note that Foord et al. [1991] indicated 70,000 carats of spessartine, but this appears to be a misprint.) Spaulding Jr. (pers. comm., 2001) also estimates the total production from the district at 40,000–50,000 carats of facet-grade spessartine. Much of the spessartine obtained by Spaulding Sr. was sold to George Ashley in Pala for cutting and distribution. Spaulding Jr. sold most of his facetable spessartine to Jonté Berlon Gems.

LOCATION AND ACCESS

The Little Three mine area is located in the Hatfield Creek valley, approximately 50 km (31 miles) northeast of metropolitan San Diego, and about 6 km (4 miles) east-northeast of the rural town of Ramona. A paved highway leads from Ramona to a private dirt road that is shared by several landowners, and public access is prohibited. The authors visited the area at various times during 1997–2001.

Lying at an altitude of 580–700 m (1,900–2,300 feet), the mining area is a scenic parched terrain with rolling hills, covered with the dense scrub vegetation typical of San Diego County's chaparral regions. The climate includes a little frost in winter, but temperatures can exceed 105°F (40.6°C) in summer. Most mining activity has taken place above ground, and the small diggings eventually become concealed by regrowth of the chaparral.

GEOLOGY

The geology of the Ramona pegmatites has been documented by numerous researchers, and it remains one of the best-studied pegmatite districts in the world (see, e.g., Merriam, 1948; Sinkankas, 1957; Simpson, 1960, 1963; Stern et al., 1986; Foord et al., 1989; Patterson, 1997; Morgan and London, 1999). The area is characterized by a complex series of intrusions that consist mostly of tonalite, with gabbro-diorite, diorite, granodiorite, and granite. These rocks have been mapped, classified, and named differently by various geologists, as compiled by Patterson (1997); they are all part of the Peninsular Ranges batholith of mid-Cretaceous age (120–100 million years [m.y.] old; Krumenacher et al., 1975). Radiometric dating of the muscovite from the Hercules mine yielded ages between 98.7 and 100 m.y. (Foord et al., 1991).

The Ramona district contains a series of pegmatite dikes in a northwest-trending belt about 4.5 km (2.8 miles) long and up to 1.6 km (1 mile) wide. Overall, the dikes strike northwest, and dip gently to moderately west to southwest. The pegmatites mined thus far typically range in thickness from 0.4 to 1.8 m (1.3 to 6 feet), and can be traced for distances up to 1,060 m (3,500 feet). In the vicinity of the Little Three mine, the dikes form locally undulating dip-slopes (i.e., erosion exposed the top surface of the pegmatites), so they were well positioned for surface mining.

Both Simpson (1960, 1963) and Stern et al. (1986) presented detailed maps of the mineralogical zoning. The dikes are typically layered, with pegmatite in the upper portion and aplite in the lower. For simplicity, these composite dikes are referred to simply as "pegmatites" throughout this article. The aplite is a fine-grained assemblage of quartz, sodic plagioclase, K-feldspar, schorl, and almandine-spessartine. The upper portion of the aplite is conspicuously banded, which miners refer to as "line rock." The coarse-grained pegmatite is formed mostly of "graphic" quartz-perthite intergrowths (Simpson, 1962) and schorl, almandinespessartine, and muscovite. Near the contact of these two units is a "pocket" zone that is characterized by relatively large euhedral crystals of quartz, feldspar, and other minerals. These minerals are often intergrown except in cavities where well-formed, free-standing crystals are attached to the pocket walls or suspended in clay. The largest, most economically important pockets in the Ramona district have yielded blue topaz, dark green tourmaline, and pale blue or pink beryl, but no spessartine (see, e.g., Foord et al., 1989).

The Ramona pegmatites show distinct differences in mineralogy and style of mineralization, even within a specific pegmatite (see, e.g., Sinkankas, 1957; Stern et al., 1986; Foord et al., 1989). Some of the pegmatites are mineralized only with spessartine (e.g., the Spaulding mine), whereas others have only topaz, beryl, and/or colored tourmaline assemblages (e.g., the A.B.C. mine); also, various assemblages may be present in the same pegmatite, but in different areas separated by several meters (e.g., the Little Three and Surprise mines). Kunz (1905) was the first to describe this mineralogical zoning. Sinkankas (1957) reported that gem spessartine has been found only in areas that do not have colored tourmaline, which was confirmed by Spaulding Jr. with continued mining.

Most of the spessartine has been mined from small pockets and a unique vuggy feldspar-rich rock with little or no quartz (see box A). This distinctive rock contains numerous small, irregularly shaped corroded cavities, and has been called "cottonball spar" or "popcorn rock" by miners, who recognized it as a good indication of spessartine mineralization (Simpson, 1960; J. Sinkankas, pers. comm., 1998). The corroded rock locally crosscuts the pocket zone and overlying pegmatite, but is not found in the underlying aplite. It is commonly located in areas of pegmatite that are jointed and fragmented with subsequent recrystallization of quartz on the fracture surfaces (J. Sinkankas, pers. comm., 1998).

Simpson (1960, 1963) noted that spessartine is found in a blue-gray clay (containing tourmaline needles) in the vugs or in larger cavities and fractures, sometimes with quartz, albite, potassium feldspar, and schorl (see also Sinkankas, 1997); rarely, it occurs with pale blue-to-pink beryl or pink apatite (L. Spaulding Jr., pers. comm., 2001). Although the spessartine in the "popcorn rock" is always corroded, the garnet crystals in the larger cavities may be either euhedral or corroded. In some places, such as in the productive open cut of

Rarity and Formation Conditions of Spessartine

What are the conditions necessary to form gem-quality spessartine in granitic pegmatites? Traces of spessartinealmandine garnet are not uncommon in these pegmatites, where they form as small (2 mm or less) fractured crystals that are intergrown with the enclosing feldspars, quartz, and schorl tourmaline. However, very few pegmatites contain facetable spessartine. At the Ramona mines, the common occurrence of gem-spessartine within vuggy, corroded feldspar-rich rock suggests its formation is related to this rock type. This appears to be the case at other gem spessartine pegmatite as well, such as Alto Mirador in northeastern Brazil (Figure A-1).

The corrosion and secondary crystallization of minerals (including spessartine) within the vuggy pegmatite at Ramona has been attributed to late-stage hydrothermal fluids that persisted after crystallization of the dikes (Sinkankas, 1957, 1997; Simpson, 1963). Such fluids evolve during the crystallization of all miarolitic (pocket-bearing) pegmatites (Černý, 2000), but only rarely does gem spessartine form. Sinkankas (1997) suggested that the hydrothermal solutions were introduced from "below" via channels created by earth movements. While such "open-system" conditions may have locally contributed to the etching and regrowth of some minerals in the pockets and vuggy rock, stable isotopic studies by Taylor et al. (1979) suggest that pocket minerals sampled from the Little Three and Hercules-Spessartite dikes crystallized under dominantly "closed-system" conditions.

The most important geochemical considerations for spessartine ($Mn_3Al_2Si_3O_{12}$) crystallization in granitic pegmatites appear to be: (1) pre-enrichment of the melt in Mn, and (2) concentration of Mn via pegmatite crystallization until the late-stage formation of gem-bearing pockets. The generation of Mn-enriched melts was mentioned briefly by London et al. (2001). Mn is not abundant in most granitic melts, and is concentrated (with other rare elements such as lithium, beryllium, and boron) in pegmatites through a process known as fractional crystallization. Mn is incompatible with the feldspars and quartz that constitute the bulk of pegmatite crystallization, so it becomes progressively concentrated during crystallization, particularly in the presence of fluorine (see, e.g., Černý et al., 1985). Together with rare elements and volatiles, Mn accumulates in the late-stage aqueous fluids that form the pegmatite pockets (see, e.g., Shigley and Kampf, 1984; Černý et al., 1985). It has been proposed that some of the Mn needed to form the gem spessartine was derived from the corrosion of previously crystallized almandine-spessartine (Sinkankas, 1997).

Why is spessartine so rarely found in the pockets? During pegmatite crystallization, the concentration of other elements (especially lithium) may stabilize minerals that consume Mn. The late-stage accumulation of Li may stop spessartine from forming in the pockets because the Mn will be preferentially incorporated into lepidolite and elbaite. At the Little Three mine in particular, Morgan and London (1999) and London et al. (2001) suggested that this mechanism halted spessartine crystallization. Likewise, the accumulation of phosphorus (P) can consume manganese through the crystallization of Mn-Fe phosphates. As expected, the important spessartine-bearing dikes in Ramona (Hercules-Spessartite and Spaulding) do not contain lepidolite, elbaite, or phosphates, but it is conceivable that traces of these minerals could be present in other pegmatites that contain enough Mn to crystallize spessartine at late stages.

The bright orange color of gem spessartine from Ramona and other localities is achievable only with low Fe contents. Iron is typically more abundant than Mn in granitic pegmatites, but it also is depleted from the melt far faster than Mn by the crystallization of mafic minerals such as schorl and biotite (Foord and Kleck, 1993; London et al., 2001). Biotite is rare or nonexistent in the gem spessartine dikes in Ramona, so schorl (which is common) is the mineral responsible for depleting Fe. Indeed, London et al. (2001) reported that Mn is incompatible in schorl, so crystallization of this tourmaline can contribute to an increasing Mn:Fe ratio in the residual pegmatitic melt-fluid system.

The rarity of commercially important pegmatitic deposits of gem spessartine is apparently due to the strong geochemical correlation between Mn and Li in granitic pegmatites (Černý et al., 1985). Spessartine is common within the inner zones of "complex" granitic pegmatites that contain lithium (Baldwin and von Knorring, 1983), although it typically is not gem quality. The crystallization of facetable spessartine requires unusual conditions favoring enrichment of Mn, but not Li (and P), as well as the depletion of Fe in the late stages of pegmatite evolution.

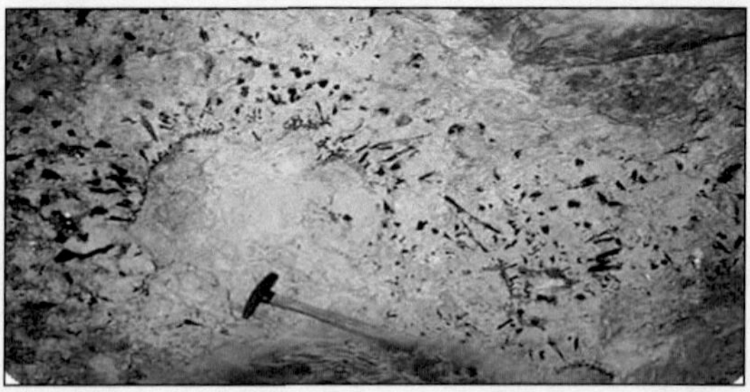

Figure A-1. Gem spessartine is commonly associated with vuggy, corroded pegmatite. At the Alto Mirador pegmatite in northeastern Brazil (see Johnson and Koivula, 1999), facetable spessartine was recovered from a gem pocket and associated corroded pegmatite in the area shown here—particularly to the left of the large feldspar crystal (marked by a partial rim of schorl) above the hammer. The quartz core zone of the pegmatite is visible to the upper right. Photo by Brendan M. Laurs.

the Spessartite mine, the feldspar-rich rock in the center of the pegmatite was altered to a 40-cm-thick layer of pink clay, in which the gems were "floating" (L. Spaulding Jr., pers. comm., 2001). Further north in this pegmatite (which was productive over about 45 m) spessartine from the Hercules mine typically formed in pockets with little or no associated corrosion of the surrounding pegmatite; here, the spessartine was not etched and wellformed mineral specimens were recovered, with the orange garnet commonly perched on cleavelandite (bladed albite) associated with schorl and smoky quartz (J. Sinkankas, pers. comm., 1998).

MINING

As is the case with most pegmatites, experience and intuition are key to determining where to dig for spessartine and other gems in the Ramona district. Most of the mining has been surficial, accomplished by "peeling" away the upper portion of the pegmatites that lie exposed on the hillside. This was initially done with a pry bar and occasional explosives. In addition to small open cuts, a few tunnels explore portions of the pegmatites that lie further underground. Most of the work done by Spaulding Jr. has been accomplished in open cuts using a bulldozer, backhoe (*Figure 10*), and a minimal amount of drilling and blasting.

The gem-bearing pockets are carefully excavated using chisels, ice picks, and wooden sticks (*Figure 11*). The pocket contents are carefully cleaned in oxalic acid to remove iron staining. Mineral specimens—rather than gem rough—are most sought after. The spessartine crystals attain sizes up to 3–4 cm and form showpieces when attached to a matrix of white cleavelandite and black schorl tourmaline (see, e.g., *Figure 12* here and *Figure 3* of Foord and Kleck, 1993). Only broken, damaged, or otherwise non-specimen-quality crystals are faceted. According to Spaulding Jr., only about 1% of the spessartine found in the pockets has been facetable.

MATERIALS AND METHODS

The Ramona study samples consisted of 14 faceted spessartines (0.47–15.93 ct) loaned by Jonté Berlon Gems, and 16 faceted (0.27–1.26 ct) and two mineral specimens (with a total of four garnets on matrix) from the GIA collection. The two matrix samples were typical of dozens of Ramona spessartine mineral specimens examined by the senior author over the past several years. Standard gemological properties were obtained on all the faceted stones. A Duplex II refractometer with a near-sodium equivalent light source was used for refractive index readings. Specific gravity was determined by the hydrostatic method,

Figure 10. Most of the spessartine production from Ramona came from the Spessartite mine, which was expanded with a backhoe. The pegmatite is dipping into the open cut from the right side. Courtesy of Louis Spaulding Jr.

Figure 11. Spessartine-bearing pockets are carefully excavated by hand, as shown here at the Spessartite mine. Courtesy of Louis Spaulding Jr.

Figure 12. Fine mineral specimens from the Ramona pegmatites are prized by collectors. The aesthetic combination of orange spessartine, black schorl, and white cleavelandite are unrivalled by spessartine specimens from other localities. Gemstones such as the fine 5.15 ct example shown here are faceted only from broken, damaged, or otherwise non-specimen-quality crystals. Photo © Harold and Erica Van Pelt

and a desk-model spectroscope was used to examine the absorption spectra. Reaction to ultraviolet radiation was viewed with four-watt longand short-wave UV lamps. Internal features and anomalous birefringence were observed with a standard gemological microscope; the latter was also viewed with a polariscope. Laser Raman microspectrometry was performed using a Renishaw 2000 Ramascope on selected faceted samples to identify mineral inclusions.

Four of the samples — ranging from light orangy yellow to medium yellow-orange, which is representative of most faceted material from Ramona (*Figure 13*) — were analyzed further. Their chemical composition was measured by electron microprobe (Cameca SX-50), and UV-Vis-NIR absorption

Figure 13. These four orangy yellow to yelloworange spessartines (0.48–1.15 ct), part of the study sample, show the range of color that is most representative of faceted material from Ramona. Photo by Maha Tannous.

spectra were obtained in the region 250–2500 nm with a Hitachi U4001 spectrophotometer. Grain mounts (one per sample) of spessartine from selected worldwide localities (i.e., Minas Gerais and Rio Grande do Norte, Brazil, and Maevatanana and the Sahatany Valley in Madagascar) were also analyzed by electron microprobe for comparison with the Ramona samples. Data for spessartine from other localities were taken from the literature.

RESULTS

Description of the Rough. The two mineral specimens consisted of gem-quality spessartine crystals on a matrix of cleavelandite feldspar; one specimen also had schorl. The spessartine was light yellowish orange to medium brownish orange with no eye-visible color zoning. The crystals were transparent in areas between fractures, some of which were partially healed. All of the spessartines on one of the specimens showed corrosion features consisting of lustrous micro-stepped surfaces (see, e.g., *Figure 14*). These etch features appear similar to those observed on spessartine from other pegmatites (see, e.g., *Figure 1* of Sinkankas and Reid, 1966). Both dodecahedral and trapezohedral faces were observed on the crystals. The dodecahedral faces showed subtle diamond-shaped growth hillocks, while the trapezohedral faces were smooth on the specimen that was not corroded.

Visual Appearance and Gemological Properties. The results of the gemological characterization are presented in *Table 1*, together with data for spessartines from other localities for comparison.

The samples ranged from light orangy yellow to dark brownish orange; most were light orangy yellow to medium yellow-orange (again, see *Figure 13*). These samples represent the range of color of gem spessartine from the Ramona

Figure 14. Corroded spessartine crystals typically show a series of lustrous micro-stepped surfaces. Photomicrograph by John I. Koivula; magnified 2×.

district. None of the stones showed eye-visible color zoning. The samples were transparent with relatively few inclusions visible with the unaided eye.

Refractive indices ranged from 1.799 to 1.808, although samples in the commercially most desirable orangy yellow to yellow-orange color range had R.I. values of 1.799–1.803. Overall, the lightest orange samples had lower R.I. values. S.G. values fell within 4.19–4.26, and showed no correlation to color or R.I. value. All stones were inert to both shortand long-wave UV radiation.

Microscopic observation revealed wavy planes of oneor two-phase (liquid-gas) inclusions in nearly all stones (see, e.g., *Figure 15*), as documented in spessartine by Gübelin and Koivula (1986), Rouse (1986), and Webster (1994). Individual inclusions within these partially healed fractures and "fingerprints" appeared as points, subparallel elongate and needle-like forms, and irregular bodies. In five samples, we saw linear arrays of tiny particles forming wisps and clouds; these were typically associated with the fingerprint-like features. Flat, isolated twophase (liquid-gas) inclusions were present in two samples (see, e.g., *Figure 16*). Three-phase (solid-liquid-gas) inclusions also were seen in a few samples. Fractures (i.e., "feathers") were rare, and showed jagged outlines.

Several samples had relatively large negative crystals and/or elongate etch channels. These channels typically showed striated surfaces and irregular angular shapes (see, e.g., *Figure 17*), although one had a rectangular cross-section that tapered lengthwise into a tube. A swarm of minute, parallel needles (probably hollow tubes) was seen in two samples. Mineral inclusions were rare. Raman analysis identified irregular-shaped, colorless inclusions with a rough texture in three samples as albite (see, e.g., *Figure 18*), and a black tapered crystal in one sample as columbite (*Figure 19*). Another sample contained

TABLE 1. Properties of spessartine from various localities.[a]

Property	Reference no. of samples	Color	Optic character	Refractive index range	Specific gravity	Internal features	Absorption spectra (nm)[g]
Ramona, California	This study/30	Orangy yellow to brownish orange	ADR	1.799–1.888	4.19–4.28	Wavy partially healed fractures, growth patterns, negative crystals, etch channels, needles/tubes, linear clouds, isolated fluid inclusions, feathers, mineral grains (albite, columbite)	Cutoff at 435–440; bands at 460, 480–485, and 520–525; 504 line and 570 band in some samples
Ramona, California[b]	Tisdall (1961)/4	Yellowish orange	nr	>1.81	4.17	Angular three-phase inclusions, "fingerprints"	Cutoff at 439; wide band at 461, wide, very strong band at 482; faint shadow at 502, faint broad band at 527
Minas Gerais, Brazil[c]	Unpublished GIA research/1	Orange	Slight ADH	1.8	4.2	Subparallel reflective stringers, irregular needles, small transparent crystals, parallel planar growth zoning	Spessartine spectrum, with very strong band at 430
Madagascar	Tisdall (1962)/nr	Orange-red, dark to pale	nr	>1.81	4.17	No three-phase inclusions; fine, close striae (like synthetic ruby)	Absorption in blue-violet, with strongest band centered at 432; intense bands centered at 461 and 484; weak bands at 585 and 527
Mozambique	Bank et al. (1988)/nr	Brownish orange	Strong ADR	>1.81	4.15–4.18	Growth zoning; feathers of film-like fluids; minerals observed as needles or platelets with high birefringance	Absorption features at 488, 422, 428, 488, 483, and 524
Namibia[d]	Koivula et al. (1993)/3; Lind et al.	Orange	nr	1.790–1.797	4.09–4.15	Angular and roiled growth structures, curved fibers, rounded crystals, dark angular crystals	Cutoff at 445, distinct diffused band above 460; stronger, wider diffused band at 480–490; weak diffused band at 525–545. Also lines at 408, 421, and 430
Namibia[e]	Johnson and Koivula (1996a)/7	Orange	SR	1.789–1.790	4.12–4.14	Quartz, fibrous amphibole, oxides (ilmenite, hematite, senaite), barite	Lines at 412, 424, 432 (edge), 462, 485, and 495
Iseyin, Nigeria	Milisenda and Zang (1999)/6, Zang et al. (1999)/9	Orange-yellow	nr	1.801–1.803	4.18–4.23	Two-phase inclusions along healed fractures, zircon with tension cracks	"Typical spessartine absorption features" plus local maxima at 469, 482, and 527
Pakistan[f]	Henn (1996)/n.r.; Johnson and Koivula (1996b)/4	Orange-red to "cherry" red; slight yellowish orange to brownish orange or red-orange	Weak ADR	1.798–1.802	4.05–4.18, 4.19–4.20	"Fingerprints," growth zoning, K-feldspar, acicular inclusions	Mn2+ bands at 408, 421, 430, 460, and 482; Fe^{2+} bands at 503, 524, and 564
Lundazi, Zambia	Johnson et al. (1999)/5 (1994)/13	Orange	nr	1.799–1.800	4.20–4.25	Typical garnet growth bands and "fingerprints"	Cutoff at 435, bands at 460, 480, and 530; weak 504 line

Footnotes for Table 1:

[a] bbreviations: ADR = anomalous double refraction, nr = not reported, SR = singly refractive. In addition, 50 spessartine-almandine garnets from Amelia, Virginia, ranging from dark red-brown to pale orange, had R.I.'s of 1.795–1.809 and S.G.'s of 4.16–4.19 (Sinkankas and Reid, 1966).

[b] See also Anderson (1961).

[c] 4.81 ct sample; see also Bank et al. (1970) and Johnson and Koivula (1999).

[d] Also known as Mandarin garnet; from the "original" mining area in northwest Namibia that was discovered in 1992; see also Koivula et al. (1993).

[e] Also known as Mandarin garnet; from a different mining site about 30 km south of the original site, discovered in 1995; see also Kammerling et al. (1995), Bank et al. (1996).

[f] Also known as Kashmirine.

[g] Determined by either spectroscopic or spectrophotometric means; values reported for this study are as observed with the spectroscope.

Figure 15. "Fingerprints" consisting of wavy planes of variously shaped two-phase (liquid-gas) inclusions were commonly observed in the Ramona spessartine. Photomicrograph by John I. Koivula; magnified 35×.

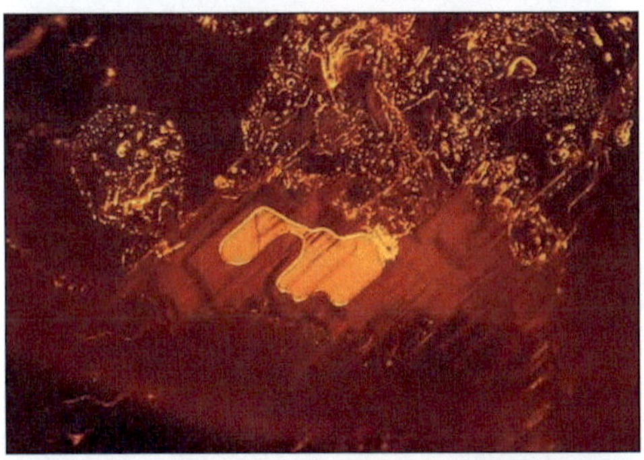

Figure 16. Flat, isolated two-phase (liquid-gas) inclusions were seen in only two spessartine samples. Note also the partially healed fractures consisting of minute two-phase inclusions in the upper part of the photo. Photomicrograph by John I. Koivula; magnified 20×.

Figure 17. This irregular-shaped etch channel in a 0.61 ct spessartine is bounded on both sides by strong growth lines. Photomicrograph by John I. Koivula; magnified 15×.

Figure 18. Raman analysis confirmed that this rough-surfaced, irregular platy inclusion is albite. Photomicrograph by John I. Koivula; magnified 15×.

Figure 19. Surrounded by a stress fracture, this black inclusion in spessartine was identified as columbite by Raman analysis. Photomicrograph by John I. Koivula; magnified 10×.

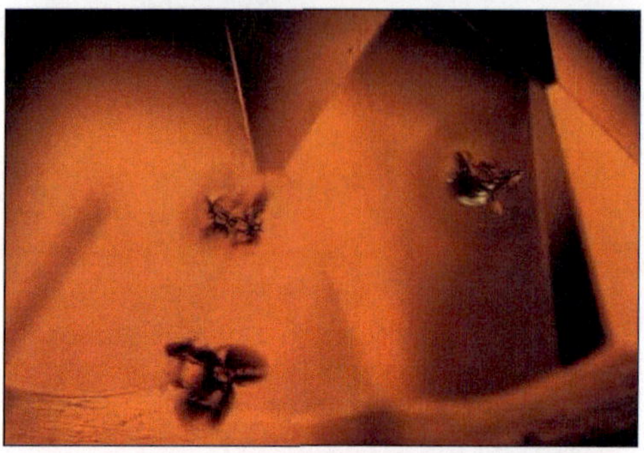

Figure 20. These three minute colorless crystals surrounded by tension fractures appear to be zircon, but they could not be identified by Raman analysis due to their size and position within the stone. Photomicrograph by John I. Koivula; magnified 20×.

Figure 21. Parallel planar growth zoning was conspicuous in a few of the spessartines, with angular and irregular growth patterns also noted. Photomicrograph by John I. Koivula; magnified 10×.

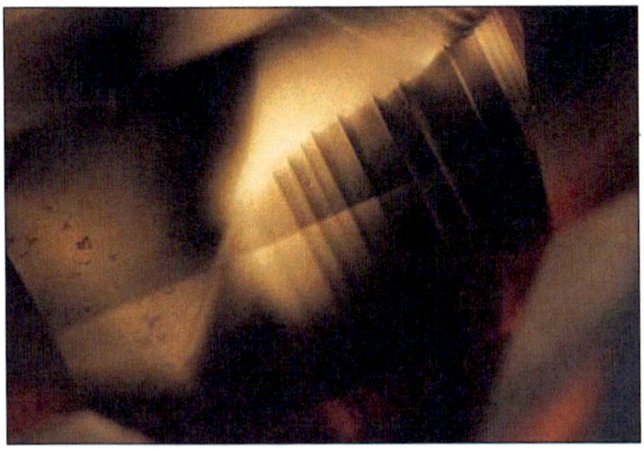

Figure 22. Anomalous birefringence was evident as irregular patterns with diffuse boundaries in all the spessartine samples. Here, planar growth zoning transects an area showing anomalous birefringence. Photomicrograph by John I. Koivula; magnified 10×.

three minute colorless crystals surrounded by tension fractures and subtle dark halos (*Figure 20*); these appeared to be zircon, but they could not be identified by Raman analysis due to their small size and position in the stone.

Most samples showed faint growth zoning, with planar, angular, and irregular patterns. Strong planar growth zoning was seen in a few samples (see, e.g., *Figure 21*), where it corresponded to sharply defined planar zones of anomalous birefringence that were visible in certain orientations. All samples displayed anomalous birefringence (see, e.g., *Figure 22*), typically as light gray areas that appeared cloud-like with diffuse boundaries, or as irregularly shaped zones that were centered around points or nodes when the samples were rotated. In some samples, patches of birefringence had a cross-hatched appearance. With the exception of the strong planar pattern mentioned above, the birefringence frequently did not show any relation to the growth features seen with plane polarized light. When viewed with the polariscope, anomalous double refraction was visible in most of the samples; the birefringence formed gray patches or linear zones with subtle to obvious intensity.

Chemical Composition and Absorption Spectroscopy. The four samples analyzed by microprobe showed a consistent decrease in manganese and increase in iron as the color varied from light orangy yellow to medium yellow-orange (Table 2). The greatest manganese concentration (40.67 wt.% MnO) was recorded in the lightest orangy yellow sample, and falls near the highest values found in the literature for gem-quality spessartine. Traces of titanium and calcium were detected in all four samples, but magnesium was below the detection limit (<0.01 wt.% MgO). The composition of these Ramona spessartines can be summarized as $Sps_{94.8-88.8}Alm_{4.7-11.0}Gro_{0.5-0.2}$.

The spectroscope revealed a cutoff at 435–440nm in all samples. Weak to moderate bands were centered at approximately 460 nm and 520–525 nm, and a strong band was present at approximately 480–485 nm. The latter band formed a region of general absorption with the 504 nm band in the darkest samples. In lighter samples, the 504 nm band was very weak or nonexistent. Another band at 570 nm could be seen in all but the lightest stones.

All four Ramona samples showed similar absorption spectra with the spectrophotometer in the UVVis-NIR range. The strongest peak was located at approximately 409 nm; on the shoulder were two additional strong peaks at approximately 421 and 430 nm (*Figure 23*). Also present in the visible range were bands at approximately 460, 483, 504, 525, and 570 nm. Within the near-infrared region were broad regions of absorption centered at about 1300 and 1700 nm (the latter being very subtle), and two bands of low intensity around 2300 nm.

DISCUSSION

For several decades, the Ramona area was the principal source of the world's high-quality spessartine gemstones. Other historic localities for gem-quality spessartine include Madagascar (Tisdall, 1962); Amelia, Virginia (Sinkankas and Reid, 1966; Sinkankas, 1968); Ceará, Brazil (Pough, 1965; Cassedanne and Cassedanne, 1977b); Minas Gerais and Rio Grande do Norte, Brazil (Bank

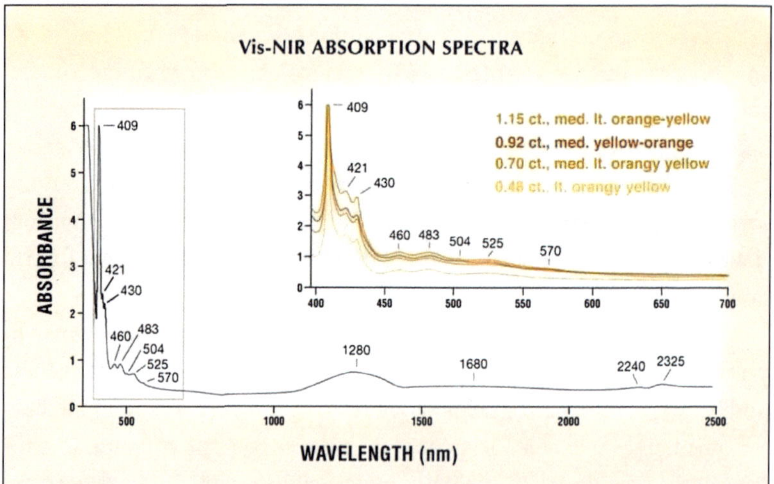

Figure 23. This UV-Vis-NIR absorption spectrum of a 0.70 ct mediumlight orangy yellow spessartine is representative of the Ramona material. It shows a dominant peak at 409 nm, with additional strong peaks at 421 and 430 nm and several smaller peaks; broad regions of absorption in the near-infrared are located at around 1300, 1700, and 2300 nm. The inset shows the absorption spectra of all four samples pictured in Figure 13 in the visible region (400–700 nm). Note the subtle increase in the strength of the 504 nm peak in the darker samples. The four samples measured 2.9–4.3 mm from table to culet; path distances of the beam could be significantly longer.

et al., 1970; Cassedanne and Cassedanne, 1977a; Johnson and Koivula, 1999; Ferreira et al., 2000); Kenya (Medenbach et al., 1978); and Sri Lanka (Noor Deen, 1984). Myanmar and India have also been reported as spessartine sources (Webster, 1994). More recently, gem spessartine has been recovered from Pakistan (Henn, 1996; Johnson and Koivula, 1996b) and Mozambique (Bank et al., 1998). Particularly fine material has been produced from Namibia (Koivula et al., 1993; Lind et al., 1994; Johnson and Koivula, 1996a), Nigeria (Milisenda and Zang, 1999; Zang et al., 1999), and Zambia (Johnson et al., 1999). All of these deposits probably have a common geologic origin (granitic pegmatites), except for those in Namibia. The gemological and chemical characteristics for spessartine from Ramona are discussed below and compared to data from other localities.

Ramona spessartine that is most likely to be seen in the gem trade is characterized by an intense orangy yellow to yellowish orange color; R.I. values of 1.799–1.803; S.G. values of 4.19–4.26; internal features consisting most commonly of wavy planes of partially healed fractures, negative crystals, and growth patterns (planar, angular, and irregular); and absorption bands at approximately 460, 480–485, 504, 520–525, and 570 nm, with a cutoff at 435–440 nm.

To the best of our knowledge, Tisdall (1961) published the only other article on the gemology of Ramona spessartines. He reported a higher R.I. value (>1.81) than we measured in our samples. However, our values are similar to those reported for spessartine by Stockton and Manson (1985: 1.798–1.808), as well as by Liddicoat (1989) and Webster (1994): 1.790–1.797 or 1.789–1.790. Our R.I. values also are consistent with those reported for other localities (see *Table 1*), with the exception of Namibian spessartine with its lower range (1.789–1.797). Also, reported values for samples from Madagascar and Mozambique were over the limits of a standard refractometer (>1.81). Sinkankas and Reid (1966) documented somewhat lower R.I. values of 1.795–1.798 for near-end-member spessartine from Amelia, Virginia, and a general increase in

R.I. with increasing iron content and darker color (e.g., approximately 1.808 for compositions around $Sps_{62}Alm_{38}$). We also recorded a general increase in R.I. with darker color. In the four chemically analyzed samples, there was a systematic correlation of R.I. with spessartine-almandine content: 1.799 to 1.802 for $Sps_{94.8}Alm_{4.7}Gro_{0.5}$ to $Sps_{88.8}Alm_{11.0}Gro_{0.2}$.

The S.G. values we obtained, 4.19–4.26, were higher than the 4.17 value determined by Tisdall (1961) using ethylene dibromide, and also higher than the general range given by Liddicoat (1989) and Webster (1994) for spessartine:

4.12–4.18 and 4.12–4.20, respectively. However, our values fall within the range Stockton and Manson (1985) determined for spessartine (4.18–4.27), and they are comparable to those reported for spessartine from some other localities (i.e., Brazil, Nigeria, and Zambia; see *Table 1*).

The features we observed with magnification are consistent with the "fingerprints" (partially healed fractures) and angular three-phase inclusions that Tisdall (1961) found in his four samples of Ramona spessartine. The negative crystals and growth zoning seen in our samples also have been previously noted (Rouse, 1986 and Liddicoat, 1970, respectively; see also *Table 1*). However, the linear clouds and inclusions of albite and columbite identified in this study have not been reported in spessartine from other localities. It was not surprising to find albite inclusions, since Ramona spessartine is commonly intergrown with albite (cleavelandite). Also, Nb-Ta oxides such as columbite are relatively common in late-stage albitic mineral assemblages in granitic pegmatites (see, e.g., Černý, 1992). Inclusions of schorl also have been reported in spessartine from Ramona (E. Gray, pers. comm., 2001), but not from other localities.

The four Ramona samples we analyzed by electron microprobe were similar in chemical composition to those from the other pegmatite localities in *Table 2* (i.e., all but Namibia), although higher calcium contents were recorded in spessartines from most of the other deposits. The highest Mn contents (i.e., spessartine component) were measured in samples from Ramona, Zambia, Brazil (Rio Grande do Norte), and Nigeria. The lowest Fe contents (i.e., almandine component) were found in material from Nigeria and Namibia. The compositions of samples analyzed from Madagascar and Pakistan are similar to one another, and show higher Ca and lower Mn than the others. The Namibian spessartine is enriched in magnesium compared to the others, reflecting its unique geologic environment within metamorphic rocks. Magnesium is not typical of granitic pegmatites (Černý et al., 1985), although traces may be derived from host rocks surrounding a pegmatite.

The visible absorption features we observed are comparable to data reported by Tisdall (1961) for Ramona spessartine. The Mn^{2+} bands (at approximately 409, 421, 430, 460, and 483 nm) recorded in our samples also are consistent with those measured by Lind et al. for spessartine from Ramona and Namibia (see also Manning, 1967; Stockton and Manson, 1985). Also, Fe^{2+}-related bands at approximately 504, 525, and 570 nm in our samples are consistent with those reported by Lind et al. and Stockton and Manson. However, additional Fe^{2+}-related bands at 610 and 690 nm recorded in Ramona spessartine by Lind et al. were not evident in the samples we analyzed (again, see *Figure 23*).

TABLE 2. Chemical composition of gem-quality spessartine from various deposits.[a]

	Ramona, California[b]				Minas Gerais, Brazil[c]	Rio Grande do Norte, Brazil[d]	Sahatany Valley Madagasgar[e]	Maevatanana, Madagasgare	Namibia	Nigeria	Pakistan	Zambia
Reference	This study	This study	This study	This study	This study	This study	This study	This study	Lind et al. (1994)	Zang et al. (1999)	Henn (1996)	This study
Description	Lt. orangy yellow, 0.48 ct, R.I.=1.799	Med.-lt. orangy yellow, 0.70 ct, R.I.=1.800	Med.-lt. orange-yellow, 1.15 ct, R.I.=1.801	Med. yellow-orange, 0.92 ct, R.I.=1.802	Orange-yellow, fragment	Lt. orangy yellow to yellowish orange, fragments	Orange-yellow, fragment	Orange-yellow, fragment	Orange	Orange-yellow	Orange-red to "cherry" red	Orangy yellow, fragment
Oxides (wt.%)												
SiO_2	36.3	36.58	36.71	36.32	36.36	36.57– 36.72	36.52	36.35	33.93–36.70	36.1	36.39	36.51
Al_2O_3	20.56	20.38	20.28	20.49	20.78	20.15–20.85	20.78	20.86	18.87–20.35	21.4	20.6	20.79
TiO_2	0.07	0.06	0.07	0.04	na	0–0.16	na	na	up to 0.11	nr	nr	nr
MnO	40.67	40.02	38.33	37.16	39.09	34.97– 41.25	35.81	35.89	37.63–42.60	40.6	36.98	40.86
MgO	< 0.01	< 0.01	< 0.01	< 0.01	< 0.01	0.44–1.48	< 0.01	<0.01	2.71–3.20	0.94	nr	<0.01
FeOf	2.06	2.89	3.62	4.64	3.04	1.03– 4.83	4.23	3.39	0.89–1.08	0.64	4.45	1.53
CaO	0.18	0.07	0.06	0.06	0.57	0.39– 0.40	1.81	2.85	0.41–0.58	0.27	1.58	0.22
Total	100.01g	100.05h	99.07	98.71	99.84	99.76–99.69	99.15	99.34	99.18–100.47	99.95	100	99.91
End members (mol.%)												
Spessartine	94.8	93.2	91.3	88.8	91.3	82.0 – 94.6	84.7	83.8	84.0–87.0	up to 94	85.3	95.8
Almandine	4.7	6.6	8.5	11	7	2.4 –10.9	9.9	7.8	1.9–2.0	1.5–11	10.1	3.5
Grossular	0.5	0.2	0.2	0.2	1.7	1.2	5.4	8.4	1.1–1.4	nr	4.6	0.7
Pyrope	—	—	—	—	—	1.8 – 6.0	—	—	9.8–12.6	3.5	—	—

[a] Abbreviations: lt.=light, med.=medium, na = not analyzed, nr = not reported. Analyses performed for this study used a Cameca SX-50 electron microprobe, with minerals or synthetic compounds as standards, an accelerating voltage of 20 kV, sample current of 15 mA, and the data correction procedure of Pouchou and Pichoir (1985). The following elements also were sought but not detected (detection limits shown are wt.%): F = 0.21, Na_2O = 0.02, P_2O_5 = 0.02, Sc_2O_3 = 0.05, TiO_2 = 0.03, V_2O_5 = 0.05, Cr_2O_3 = 0.04, ZnO = 0.09, SrO = 0.02, Y_2O_3 = 0.07, SnO_2 = 0.05. All other analyses were by electron microprobe, except for the Nigerian spessartine (by SEM-EDS). In addition, 50 electron microprobe analyses of spessartine from the Amelia, Virginia, yielded $Sps_{50}Alm_{43}Gro_7$ to $Sps_{95}Alm_3Gro_2$ (Sinkankas and Reid, 1966). Majorand trace-element data for spessartine-almandine garnets from several localities is presented in Taylor (1998).

[b] Chemical data on Ramona spessartine is found in several publications. Taylor et al. (1979) determined an oxygen isotope value (d18O) of +5.4‰ for a pocket crystal from the Hercules-Spessartite dike. In a sample containing 3.0% Fe, Foord et al. (1989) recorded the following trace elements by emission spectrometry: 0.07% Ca, 0.05% Ti, 0.0015% Mg, 70 ppm Sc, 70 ppm Ge, 150 ppm Y, 50 ppm Zr, 30 ppm Nb, 70 ppm Gd, 50 ppm Dy, and 3 ppm Yb. Taylor (1998, p. 37) recorded average values of 0.39 wt.% P_2O_5 and 0.17 wt.% TiO_2, and no Zr or F, in two electron microprobe analyses of a sample with an average composition of $Sps_{83.9}Alm_{14.7}Gro_{1.4}$; additional trace elements were measured by direct-coupled plasma emission spectrometry (ppm)—Li = 21, Nb = 83.5, Cs = 1.5, Ta = 27.5, Pr = 0.3, Nd = 1.8, Eu = 0.5, Gd = 1.5, Tb = 0.3, and Dy = 6.1. See also Foord and Kleck (1993).

[c] See also Bank et al. (1970).

[d] Range from 21 analyses of three samples from the Alto Mirador mine.

[e] See also von Knorring and Baldwin (1969) and Schmetzer et al. (2001).

[f] All iron is reported as FeO.

[g] Also contained 0.17 wt.% P_2O_5.

[h] Also contained 0.05 wt.% P_2O_5.

Overall, the absorption spectra of the four Ramona samples were similar, as expected from their relatively narrow range in color and composition (again, see *Figure 13* and *Table 2*). The spectra were dominated by an intense Mn^{2+} absorption at approximately 409 nm (again, see *Figure 23*); this peak is actually the result of combined strong absorptions at 407 and 411 nm (Stockton and Manson, 1985). The darker samples showed a subtle increase in the strength of the 504 and 570 nm peaks, as expected from the spectroscope and chemistry results. The Mn^{2+} peaks at approximately 430, 460, and 483 nm in the darkest sample were less defined than in the other samples.

The broad regions of absorption in the nearinfrared region (at about 1300, 1700, and 2300 nm) may be due to traces of hydroxyl groups (compare to Lind et al., 1994). Hydrous components in garnet are most easily detected in the infrared spectra (as absorption bands in the 3600 cm-1 region), which were not collected in this study; traces of hydrogen (expressed as 0.038 wt.% H_2O) were found in Ramona spessartine by Rossman et al. (1988) by nuclear reaction analysis.

According to Fritsch and Rossman (1987), the orange color of spessartine is caused by Mn^{2+} in a site with distorted cubic coordination. Iron causes the intrinsic yellow-orange color of spessartine to darken into red and brown hues (Sinkankas and Reid, 1966). The orangy yellow to brownish orange color seen in Ramona spessartine is due to a combination of absorptions from Mn^{2+} and Fe^{2+}.

Identification. Our review of the literature revealed no report of the growth of gem-quality spessartine in the laboratory, and we know of no treatments done to the natural material. Spessartine may resemble orange grossular, and at one time the Ramona material was thought to be hessonite (Sinkankas, 1959); it was called "hyacinth" (a varietal name for orange grossular) by the locals in the early years of mining (L. Spaulding Jr., pers. comm., 2001). Stockton and Manson (1985) studied three types of reddish orange garnet that could be confused with one another—"malaya," grossular, and spessartinealmandine—and found they could be separated by their R.I. values and spectroscope spectra.

Spessartine also may resemble zircon, sphene, topaz, and citrine (Liddicoat, 1989), but these can be separated easily with standard gemological testing. Federman (1994) noted the similarity between spessartine and Mexican fire opal, but also pointed out their obvious gemological differences. Spessartine also resembles the rare gem clinohumite, but the two minerals can be readily separated on the basis of optical properties and UV fluorescence (Hyrsl, 2001).

CONCLUSION

The historic Little Three mining area in Ramona, California, was the principal source of gem-quality spessartine for most of the 20th century. Most of the spessartine production occurred sporadically from 1956 to 1994. An estimated total of 40,000–50,000 carats of facetable rough was mined chiefly from the Hercules-Spessartite and the Spaulding dikes, within vuggy corroded pegmatite and small "pockets."

Ramona spessartines characteristically are a bright orangy yellow to yellowish orange, with few (if any) eye-visible inclusions. Internal features such as wavy planes of partially healed fractures, anomalous birefringence, and growth patterns are comparable to spessartine from other localities, but linear clouds, inclusions of albite and columbite, and the common occurrence of negative crystals in the study samples appear unique to Ramona. Four analyzed samples of the characteristic color range were spessartine rich: $Sps_{94.8}Alm_{4.7}Gro_{0.5}$ to $Sps_{88.8}Alm_{11.0}Gro_{0.2}$. Absorption spectra showed corresponding strong peaks for Mn^{2+} and subordinate Fe^{2+} features.

Because the most accessible portions of the Ramona spessartine-bearing pegmatites have been mined, a significant investment in time and work will be necessary before commercial production can resume. Also, in recent years, the Ramona locality has been overshadowed by abundant production of spessartines from Namibia and Nigeria. Nevertheless, the price of fine spessartine (as reported by *The Guide*) appears to be stable, and the highestquality material is gradually increasing in value. With the greater popularity of spessartine in both high-end jewelry and mass-produced goods (Henricus, 2000), the stone is in strong demand by the gem trade. Certainly any future production of Ramona spessartine will find a ready market.

ABOUT THE AUTHORS

Mr. Laurs (blaurs@gia.edu) is senior editor of *Gems & Gemology* at GIA in Carlsbad, California. Ms. Knox is a research consultant at the University of California at San Diego, California.

ACKNOWLEDGMENTS:

The authors thank E. "Buzz" Gray (Missoula, Montana) and Bernadine Johnston (Jonté Berlon Gems, Missoula, Montana) for loaning samples and providing helpful information. Louis B. Spaulding Jr. is thanked for extensive interviews and access to the mine site. John Koivula performed photomicrography and some of the gemological testing at the GIA Gem Trade Laboratory in Carlsbad.

Also at GIA, Shane Elen obtained UV-VisNIR spectra, and together with Sam Muhlmeister provided helpful assistance with Raman analyses. Dr. Frank Hawthorne, Ron Chapman, and Kim Tait (all at the University of Manitoba, Winnipeg, Canada) performed electron microprobe analyses. Spessartine samples from various worldwide localities (indicated in parentheses here) were provided for microprobe analysis by: Brian Cook of Nature's Geometry, Graton, California (Rio Grande do Norte, Brazil); Tom Cushman of Allerton Cushman & Co., Sun Valley, Idaho (Madagascar); James Dzurus of James Dzurus & Co., Franklin, North Carolina (Minas Gerais, Brazil); and Marc Sarosi of Marc Sarosi Co., Mill Valley, California (Zambia). Stuart Robertson (Gemworld International, Northbrook, Illinois) provided information on spessartine pricing. John Sinkankas provided useful correspondence and discussion. Dr. David London reviewed box A. BL appreciates the inspiration from the late Eugene Foord, and thanks Jeffrey Patterson (University of Calgary, Alberta, Canada) for initial field excursions to Ramona. We appreciate the assistance of Harold and Erica Van Pelt, Jeff Scovil, Bernadine Johnston, Buzz Gray, John Sinkankas, Cal and Kerith Graeber, and Maha Tannous in supplying photographs.

REFERENCES

Anderson B.W. (1961) Letter to the editor: Spessartine garnets. *The Gemmologist*, Vol. 30, No. 359, pp. 108–109.

Baldwin J.R., von Knorring O.V. (1983) Compositional range of Mngarnet in zoned granitic pegmatites. *Canadian Mineralogist*, Vol. 21, pp. 683–688.

Bank H., Berdesinski W., Ottemann J. (1970) Orangeroter Spessartin aus Brasilien. *Zeitschrift der Deutschen Gemmologischen Gesellschaft*, Vol. 19, No. 3/4, pp. 123–127.

Bank H., Henn U., Milisenda C.C. (1996) Gemmological News: Spessartite garnet from Namibia. *Gemmologie: Zeitschrift der Deutschen Gemmologischen Gesellschaft*, Vol. 45, No. 2, p. 43. Bank

H., Henn U., Milisenda C.C. (1998) Gemmological News: Garnets from Mozambique. *Gemmologie: Zeitschrift der Deutschen Gemmologischen Gesellschaft*, Vol. 47, No. 2, p. 61.

Cassedanne J.-P., Cassedanne J.-O. (1977a) Les grenates de Poaia / Brésil. *Revue de Gemmologie a.f.g.*, No. 52, pp. 2–4.

Cassedanne J.-P., Cassedanne J.-O. (1977b) Les grenates de Poço dos Cavalos (Ceará–Brésil). *Revue de Gemmologie a.f.g.*, No. 62, pp. 5–8.

Černý (1992) Geochemical and petrogenetic features of mineraliza-

tion in rare-element granitic pegmatites in the light of current research. *Applied Geochemistry*, Vol. 7, pp. 393–416.

Černý P. (2000) Constitution, petrology, affiliations and categories of miarolitic pegmatites. *Memorie della Società Italiana di Scienze Naturali e del Museo Civico di Storia Naturale di Milano*, Vol. 30, pp. 5–12 [http//www.scienzenaturali.com/redazionale/cerny.pdf].

Černý P., Meintzer R.E., Anderson A.J. (1985) Extreme fractionation in rare-element granitic pegmatites: Selected examples of data and mechanisms. *Canadian Mineralogist*, Vol. 23., pp. 381–421.

Federman D. (1994) Gem profile—Mandarin garnet: Great balls of fire. *Modern Jeweler*, Vol. 93, No. 1, pp. 39–40.

Ferreira A.C.M., Ferreira J.A.M., Tavares E J. (2000) As espessartitas do Alto Mirador, Rio Grande do Norte. *I Sympósio Brasileiro de Tratamento e Caracterização de Gemas*, Ouro Preto, Minas Gerais, Brazil, p. 28.

Foord E.E., Kleck W.D. (1993) Garnets in the gem-bearing pegmatite dikes of San Diego County, California. *Mineralogical Record*, Vol. 24, No. 1, pp. 66–67.

Foord E.E., Spaulding L.B., Mason R.A., Martin R.F. (1989) Mineralogy and paragenesis of the Little Three mine pegmatites, Ramona district, San Diego County, California. *Mineralogical Record*, Vol. 20, No. 2, pp. 101–127.

Foord E.E., London D., Kampf A.R., Shigley J.E., Snee L.W. (1991) Gem-bearing pegmatites of San Diego County, California. In

M.J. Walawender and B.B. Hanan, Eds., *Geological Excursions in Southern California and Mexico*, field trip guidebook of the 1991 Annual Meeting of the Geological Society of America, San Diego, California, October 21–24, 20 pp.

Fritsch E., Rossman G.R. (1987) An update on color in gems. Part 1: Introduction and colors caused by dispersed metal ions. *Gems & Gemology*, Vol. 23, No. 3, pp. 126–139.

Gübelin E.J., Koivula J.I. (1986) *Photoatlas of Inclusions in Gemstones*. ABC Edition, Zurich, Switzerland.

Henn U. (1996) Spessartites from Pakistan. *Gemmologie: Zeitschrift der Deutschen Gemmologischen Gesellschaft*, Vol. 45, No. 2, pp. 93–94.

Henricus J. (2000) Spessartite garnet gains market acceptance. *Jewellery News Asia*, No. 191, pp. 58, 60, 62.

Hyrsl J. (2001) Gem News International: Spinel with clinohumite from Mahenge, Tanzania. *Gems & Gemology*, Vol. 37, No. 2, pp. 144–145.

Johnson M.L., Koivula J.I., Eds. (1996a) Gem News: Update on Namibian spessartine. *Gems & Gemology*, Vol. 32, No. 1, pp. 56–57.

Johnson M.L., Koivula J.I., Eds. (1996b) Gem News: Spessartine from Pakistan. *Gems & Gemology*, Vol. 32, No. 3, pp. 218–219.

Johnson M.L., Koivula J.I., Eds. (1999) Gem News: New finds of spessartine in Brazil. *Gems & Gemology*, Vol. 35, No. 1, p. 55.

Johnson M.L., Koivula J.I., McClure S.F., DeGhionno D., Eds. (1999) Gem News: Spessartine from Zambia. *Gems & Gemology*, Vol. 35, No. 4, p. 217.

Kammerling R.C., Koivula J.I., Fritsch E., Eds. (1995) Gem News: Spessartine garnet from Namibia. *Gems & Gemology*, Vol. 31, No. 2, p. 134.

Koivula J.I., Kammerling R.C., Eds. (1988) Gem News: Garnet. *Gems & Gemology*, Vol. 24, No. 1, pp. 57–58.

Koivula J.I., Kammerling R.C., Fritsch E., Eds. (1993) Gem News: Spessartine garnet from Africa. *Gems & Gemology*, Vol. 29, No. 1, pp. 61–62.

Krumenacher D., Gastil R.G., Bushee J., Doupont J. (1975) K-Ar apparent ages, Peninsular Ranges batholith, southern California and Baja California. *Geological Society of America Bulletin*, Vol. 86, pp. 760–768.

Kunz G.F. (1905) *Gems, Jewelers' Materials, and Ornamental Stones of California*. California State Mining Bureau, Bulletin No. 37, 171 pp.

LeMenager C.R. (1989) *Ramona and Round About*. Eagle Peak Publishing Co., Ramona, CA.

Liddicoat R.T., Jr. (1970) Developments and highlights at GIA's lab in Los Angeles: Unusual spessartite garnet. *Gems & Gemology*, Vol. 13, No. 6, p. 189.

Liddicoat R.T., Jr. (1989) *Handbook of Gem Identification*. Gemological Institute of America, Santa Monica, California.

Lind Th., Henn U., Bank H. (1994) Leuchtend orangefarbige Spessartine aus einem neuen Verkommen in Namibia. *Zeitschrift der Deutschen Gemmologischen Gesellschaft*, Vol. 43, No. 1/2, pp. 39–47.

London D., Evensen J.M., Fritz E., Icenhower J.P., Morgan G.B. IV, Wolf M.B. (2001) Enrichment and accommodation of manganese in granite-pegmatite systems. In *Eleventh Annual V.M. Goldschmidt Conference*, Abstract #3369. LPI Contribution No. 1088, Lunar and Planetary Institute, Houston (CD-ROM).

Manning P.G. (1967) The optical absorption spectra of the garnets almandine-pyrope, pyrope and spessartine and some structural interpretations of mineralogical significance. *Canadian Mineralogist*, Vol. 9, No. 2, pp. 237–251.

Medenbach V.O., Schmetzer K., Krupp H. (1978) Spessartin aus den Taita Hills, Kenia. *Der Aufschluss*, Vol. 29, pp. 275–276.

Merriam R. (1948) Orbicular structures in the aplite dikes near Ramona, California. *American Journal of Science*, Vol. 246, No. 3, pp. 129–137.

Milisenda C.C., Zang J. (1999) Spessartine aus Nigeria. *Gemmologie: Zeitschrift der Deutschen Gemmologischen Gesellschaft*, Vol. 48, No. 4, pp. 223–226.

Morgan G.B. VI, London D. (1999) Crystallization of the Little Three layered pegmatite-aplite dike, Ramona district, California. *Contributions to Mineralogy and Petrology*, Vol. 136, pp. 310– 330.

Noor Deen M. (1984) Rare gemstones of Sri Lanka. *Lapidary Journal*, Vol. 38, No. 1, pp. 238–240.

Patterson J.E. (1997) The Little 3 Mine "Plug": Evidence for a Zoned Plutonic Intrusion and a Plausible Parent Source of the Ramona Pegmatite District, Ramona, San Diego County, California. Bachelor's thesis, University of Arizona, Tucson, AZ, 152 pp.

Pouchou J.-L., Pichoir F. (1985) "PAP" f(rZ) procedure for improved quantitative microanalysis. In J.P. Armstrong, Ed., *Microbeam Analysis*, San Francisco Press, San Francisco, CA, pp. 104–106.

Pough F.H. (1965) Rare faceting minerals—Garnets. *Lapidary Journal*, Vol. 19, No. 5, pp. 564–568.

Rossman G.R., Rauch F., Livi R., Tombrello T.A., Shi C.R., Zhou Z.Y. (1988) Nuclear reaction analysis of hydrogen in almandine, pyrope, and spessartite garnets. *Neues Jahrbuch fur Mineralogie, Monatshefte*, No. 4, pp. 172–178.

Rouse J.D. (1986) Chapter 6—Spessartite. *Garnet*, Ed. by P.G. Read, Butterworths Gem Books, London.

San Diego Mining Co. (2001) Hercules mine, Hatfield Creek, Ramona, San Diego County, California, http://www.socalgem.com/ sdmc/hercules/parthenon.htm [date accessed: August 3, 2001].

Schmetzer K., Hainschwang T., Kiefert L., Bernhardt H.J. (2001) Pink to pinkish orange malaya garnets from Madagascar. *Gems & Gemology*, Vol. 37, No. 4, pp. 296–308.

Shigley J.E., Kampf A.R. (1984) Gem-bearing pegmatites: A review. *Gems & Gemology*, Vol. 20, No. 2, pp. 64–77.

Simpson D.R. (1960) Geology of the Ramona pegmatites, San Diego County, California. Ph.D. dissertation, California Institute of Technology, Pasadena, CA.

Simpson D.R. (1962) Graphic granite from the Ramona pegmatite district, California. *American Mineralogist*, Vol. 47, pp. 1123–1138.

Simpson D.R. (1963) Geology of the central part of the Ramona pegmatite district—San Diego County, California. *Short Contributions to California Geology*, Special Report 86, pp. 3–23. Sinkankas J. (1957) Recent gem mining at Ramona—San Diego County, California. *Gems & Gemology*, Vol. 8, No. 12, pp.367– 373.

Sinkankas J. (1959) *Gemstones of North America*, Vol. 1. Van Nostrand Reinhold, New York, pp. 287–289.

Sinkankas J. (1967) Notes on some minerals from San Diego County, California. *Gems and Minerals*, No. 363, pp. 34–35.

Sinkankas J. (1968) Classic mineral occurrences: I. Geology and mineralogy of the Rutherford pegmatites, Amelia, Virginia. *American Mineralogist*, Vol. 53, pp. 373–405.

Sinkankas J. (1976) *Gemstones of North America—In Two Volumes*, Vol. 2. Van Nostrand Reinhold, New York.

Sinkankas J. (1997) *Gemstones of North America*, Vol. 3. Geoscience Press, Tucson, AZ.

Sinkankas J., Reid A. (1966) Colour-composition relationship in spessartine from Amelia, Virginia. *Journal of Gemmology*, Vol. 10, No. 4, pp. 125–134.

Spaulding L.B. (1977) Twenty-five years of mining in the Little Three pegmatites near Ramona, California. *Crystal Growth and Habit*, 3rd Mineralogical Society of America—Friends of Mineralogy Symposium, February 13–14, Tucson, AZ.

Stern L.A., Brown G.E. Jr., Bird D.K., Jahns R.H. (1986) Mineralogy and geochemical evolution of the Little Three pegmatite-aplite layered intrusive, Ramona, California. *American Mineralogist*, Vol. 71, pp. 406–427.

Sterrett D.B. (1908) Precious stones. In *Mineral Resources of the United States—Part 2, Non-Metallic Products*, calendar year 1907, U.S. Bureau of Mines Minerals Yearbook, pp. 795–842.

Stockton C.M., Manson D.V. (1985) A proposed new classification for gem-quality garnets. *Gems & Gemology*, Vol. 21, No. 4, pp. 205–218.

Taylor B.E., Foord E.E., Friedrichsen H. (1979) Stable isotope and fluid inclusion studies of gem-bearing granitic pegmatite-aplite dikes, San Diego Co., California. *Contributions to Mineralogy and Petrology*, Vol. 68, pp. 187–205.

Taylor L.S. (1998) Trace Element Chemistry of Pegmatitic Garnets. M.Sc. thesis, University of New Orleans, LA.

Tisdall F.S.H. (1961) Spessartite garnet from California. *The Gemmologist*, Vol. 30, No. 357, pp. 61–62.

Tisdall F.S.H. (1962) Spessartites from Madagascar. *The Gemmologist*, Vol. 31, No. 372, pp. 124–125.

Tucker W.B. (1925) San Diego County. In *Mining in California: 21st Report of the State Mineralogist*, pp. 328–382.

Tucker W.B., Reed C.H. (1939) Mineral resources of San Diego County, California. *California Journal of Mines and Geology*, Quarterly Chapter of State Mineralogist's Report, Vol. 35, No. 1, pp. 8–55.

von Knorring O., Baldwin J.R. (1969) The geochemistry of spessartine garnets from African pegmatites. *Thirteenth Annual Report of the Research Institute of African Geology*, Department of Earth Sciences, University of Leeds, England, pp. 32–34.

Weber F.H., Jr. (1963) *Geology and Mineral Resources of San Diego County, California*. California Department of Mines and Geology, County Report 3, 309 pp. (Gem Minerals, pp. 79–115).

Webster R. (1994) *Gems: Their Sources, Descriptions and Identification*, 5th ed., revised by P.G. Read. Butterworth Heinemann, London.

Zang J., Schoder K., Luhn M. (1999) Gem News: Spessartine from Nigeria. *Gems & Gemology*, Vol. 35, No. 4, p. 216

Selected Geological and Historical Aspects of the Julian Gold Mining District
September 1987

SAN DIEGO ASSOCIATION OF GEOLOGISTS

FIELD TRIP LEADERS

George Copenhaver, Geocon, Inc.

Ron Kofron, Kleinfelder & Associates

FIELD TRIP PLANNING COMMITTEE

John Franklin and Jim Glaze

SDAG EXECUTIVE COMMITTEE 1987

John Franklin, *Owen Geotechnical, Chairperson*

Jim Glaze, *American Geotechnical, Vice Chairperson*

James R. Evans, *Engineering Geology Consultant, Secretary*

Diane Murbach, *Murbach Geotechnical, Treasurer*

Curtis Burdett, *So. Calif. Soil & Testing, Inc., Publications Coordinator*

ACKNOWLEDGEMENTS

Thanks to: James R. Evans, consulting geologist, who suggested the topic and initiated the idea of reviewing local mining geology and history. Joe Johanek, owner of the Keystone claims, for permission to trespass and tour property, and his help in locating mine owners in the Julian area and other areas. George Vrable, owner of the Warlock Mine, for permission to trespass and tour properties. Ken Donlan, owner of the Chariot Mine, for his hospitality and helpfulness in tours and information on the Chariot Mine and Mill. George Hernington, owner of the Cold Beef Mine, for his hospitality and helpfulness in review of his experiences in the Julian district. Owners of the Eagle and High Peak Mines for their helpfulness and cooperation in the mine tours. George Copenhaver and Ron Kofron for their willingness to share their knowledge of the Julian District and to help organize this field trip. American Geotechnical for their help in typing and preparing this road log.

INTRODUCTION

The Julian gold mining district was significant in its initial production of new wealth (and the hope for new wealth) for the early discoverers of a number of gold deposits. Some of which will be visited on this field trip. These early discoveries were made during hard times after the Civil War and largely by Southerners or ex-Confederate soldiers who had already settled in the area (Pourade, 1964). Despite the prejudice against the Southerners by largely northern-based mining investors, the Julian-Banner-Cuyamaca miners produced a surprisingly large amount of gold—over 219,000 troy ounces, mostly prior to 1900. This ranks as one of the principal gold-producing districts of the U.S. (Koschmann and Bergandahl, 1968).

The failure of a plan in 1874 establishing a terminus for a transcontinental railroad in San Diego (which initially attracted many settlers), still-fresh anti-Southerner feelings by potential investors, and a nationwide depression, all combined to hamper development of a relatively untapped resource. As a recent study (Kofron, 1984) pointed out, "Political considerations [still] pose the largest obstacle to large-scale mining in the Julian district."

Politics could possibly work in favor of a new era of precious metals mining in San Diego County. After all, the San Diego County Health Department recently cooperated with our neighbor Imperial County in the mitigation of potential hazardous waste problems associated with the new, and very successful, Mesquite gold mine near Ogilby. This mine has been approved and permitted and is presently producing over 100,000 troy ounces of gold per year. New technology in ore and waste treatment can benefit both small and large-scale mining operations. Another environmentally sensitive area is Napa County, California, near the heart of vineyard and fruit growing country. A major gold mine, called the McLaughlin Mine, has been approved and permitted, and is producing over 200,000 troy ounces of gold per year. The mine utilizes a unique system designed to not only remove gold from the ore, but also remove mercury which is naturally abundant in the area, and constituted a potential natural hazard. The new mines recovery s stem is actually cleaning up the environment and meets the California Department of Health Services tough tests for a non-hazardous classification. The McLaughlin tailings facility is classed as a "zero discharge system" with no consequential discharge of mill process water or effluent to the environment.

The point of this field trip is not just that the Julian gold district "is history," but that we may have a potential resource which should be studied and understood from both geologic and environmental perspectives.

George Copenhaver

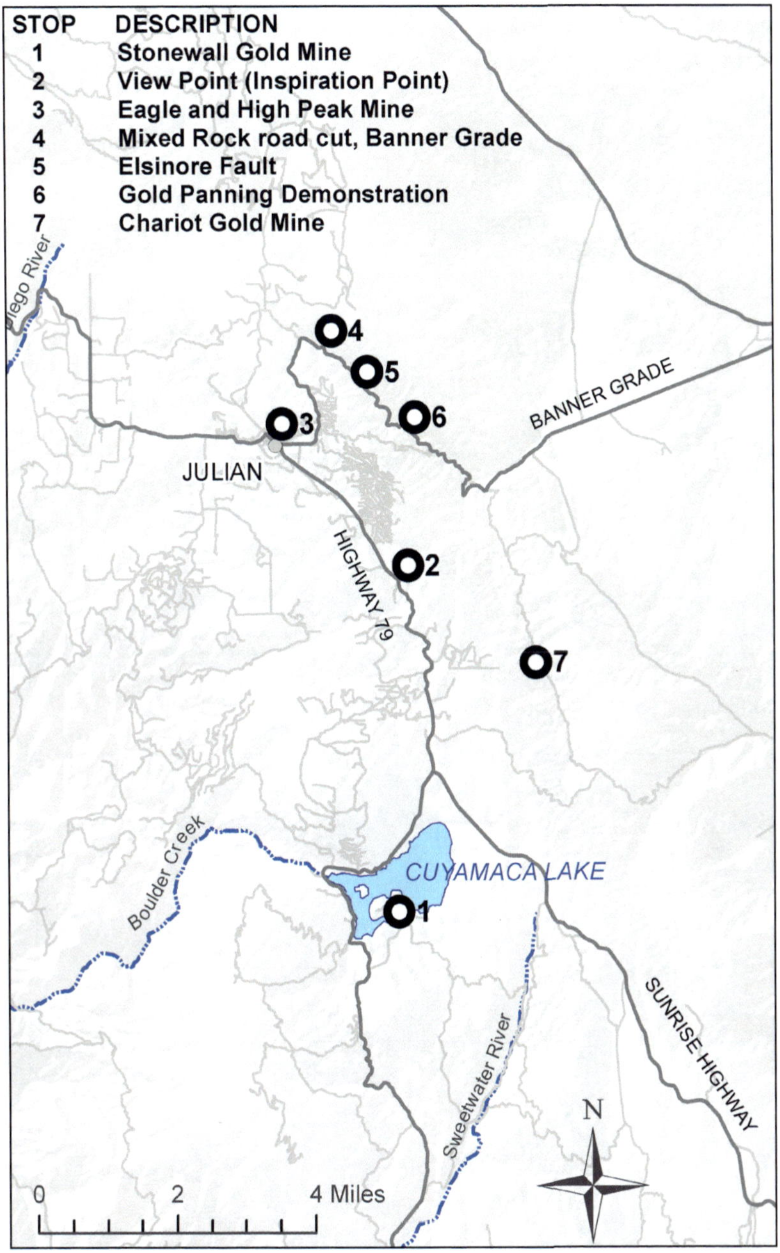

Figure 1. Map of Field Trip Stops 1–7.

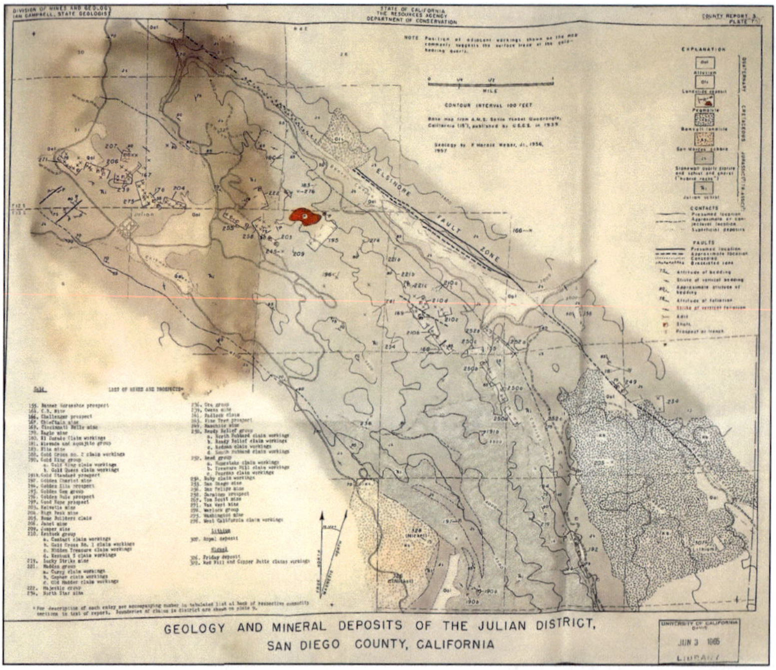

Figure 2. Geology and Mineral Deposits of the Julian District, San Diego County, California. Portion of Plate 7, County Report 3 (Weber, 1963).

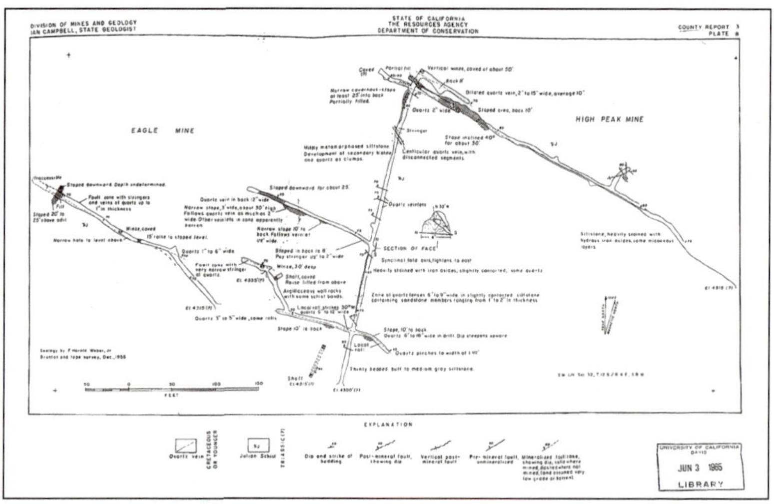

Figure 3. Map of Underground Workings of the Eagle and High Peak Mines Showing Geology and Typical Vein System of the Julian District, San Diego County, California. Portion of Plate 8 (Weber, 1963). A digital copy of this paper, including high-resolution versions of Figures 2 and 3, is available at the SDAG website, www.sandiegogeologists.org.

0.0 miles Starting point of trip is Leighton and Associates' office located at 4393 Viewridge, Suite D, San Diego.

Note Zero odometer at corner of Balboa Avenue and Viewridge.
Depart Leighton & Associates office. North on Viewridge
East on Balboa Avenue
South on Interstate Highway 5
East on Interstate Highway 8 (toward El Centro)

35.9 miles Exit north on State Highway 79 (toward Julian)

38.1 miles Left, continuing north on State Highway 79

48.2 miles Right turn (toward Stonewall Mine)

49.0 miles Remnants of tailings can be seen to the lower left of the topographic saddle; mill tailings were washed into the grass valleys downstream from this point, but apparently were not sterile enough to prevent future regrowth of grasses and herbs. It should be noted these tailings were also secondarily treated by cyanide leaching, which removed at least another 10,000 troy ounces from just the waste material.

To our right, steeply dipping foliated rocks of the metasediments hosting the massive quartz vein ore body of the Stonewall Mine can be seen cropping out. Extensions, or lenses of the quartz ore could not be found in theses rocks, either to the north or south of the main discovery site (in the saddle). The probable cause for this is either lateral pinching-out of the vein, and/or plunging structural traps, causing a vertical locus of ore-bearing quartz deposits.

49.3 miles **STOP 1**: Stonewall Mine Gold was discovered here in 1870 by William Skidmore. Skidmore was searching for a stray mule when he encountered an outcrop of gold bearing quartz. He staked a claim and started mining. The mine came to be known as the Stonewall Mine. In a short time the mine was purchased by Almon Frary. Frary mined the Stonewall for 1871 to approximately 1875. There was a production lull and various changes in ownership the next 11 years. Then from 1886 to 1893, the mine was in full swing under the ownership of Robert Waterman, Governor of California. The mine shafts were deepened from 155 feet to 600 feet. The most productive period for the mine was from 1888 to 1891 during which time over $900,000 worth of gold was extracted. By 1883, the mine was played out and the operation was stopped. Operations resumed again in 1889 when a cyanide reduction plant processed tailings obtaining approximately $200,000 worth of gold (Weber, 1963).

The Stonewall Mine was the most productive gold mine in San Diego County (but not in the Julian district). It produced over $2,000,000 in gold. The main gold shaft is 630 feet deep and was abandoned in 1926. The gold was found in massive bull quartz veins within the Jura-Triassic-aged Julian Schist and quartz diorite. The quartz veins followed major shear zones of the area. The ore body strikes northwestward and dips steeply toward the southwest (Merrill, 1914).

Little is directly known about the geology of the Stonewall Mine as the early reports do not employ any of the modern language or theory with respect to gold deposits, and the shaft has been completely inaccessible since the 1920s. However, descriptions of the quartz veins indicate that it was soft and "decomposed." As learned earth scientists, you know that it is very difficult to "decompose" quartz except by mechanical crushing. This description indicates that major shear zones active since Cretaceous time, host the quartz veins.

Exit Park.

50.4 miles Right turn on Highway 79, continue north

51.4 miles Cuyamaca Reservoir is on the right. According to Merriam, 1951, rocks immediately adjacent the road towards the north end of the reservoir are mapped as mixed rocks (Jurassic-aged quartz diorite and schist).

56.9 miles Red material on the right is lateritic or ferralitic soil derived from oxidation of sulfide (pyrrhotite-pentlandite)-rich gabbroic and ultramafic rocks of the San Marcos Gabbro[1]. Small tonnage nickel-cobalt ore bodies were mined near here in the early 1900s.

57.7 miles **STOP 2**: Right turn at View Point. The view point, according to

1 The San Marcos Gabbro may also be referred to in some texts as the Cuyamaca Gabbro.

Merriam, is located at the contact of the Cretaceous-aged San Marcos Gabbro, Jurassic-aged mixed rocks and Jura-Triassic-aged Julian Schist. Looking northeast, the Julian Schist extends downhill from the view point. This side (southwest) of the town of Banner, the schist contacts Merriam's mixed rocks. The Elsinore Fault can be seen trending in a northwest-southeasterly direction through the town of Banner. The gold mineralization occurred within the Julian Schist, generally closer to its northeastern contact with the mixed rocks.

58.0 miles Back to Highway 79, turn right continue to Julian.

59.5 miles Outcrops of Merriam's mixed rocks on right (quartz diorite and schist).

60.5 miles Left turn to Julian.

60.6 miles Right turn on road just past real estate office. Continue straight on this road; follow signs to the Eagle and High Peak Mines.

61.0 miles **STOP 3**: Eagle and High Peak Mine. Groups of approximately 15 will be taken through the mine.

Lunch may be here at the mine or at our next stop on Banner Grade depending upon time and traffic. According to CDMG County Report 3, the geology for the two mines is as follows:

Eagle Mine. "One west-northwest trending claim that adjoins the south-southwest side of the High Peak claim. The claim was located in 1870; it is unpatented. Principal working is a north-northeast trending crosscut adit 358 ft. long. This working intersects the northwest-trending drift adit of the High Peak mine. Appended to the crosscut adit are 3 west-northwest trending drifts which from the south- southeast to north-northeast are 143, 37, and 157 ft. long. About 170 ft. west of the portal of the crosscut adit, a 260 ft. drift adit extends northwestward on the same vein explored by the 37 ft. drift. Ore was mined from narrow stopes in the drifts and drift adit. Total value of production was estimated by Donnelly (1934) to be between $25,000 and $50,000. Mine operated during early 1870s, then mine and/or mill operated nearly continuously on a small scale from late 1880s to 1939. Operated from 1925 to 1939 by Sid Dodge of Julian. (Crawford 96:335; Donnelly 34:350, 352, 362, pl.4; Eric 48:320; Goodyear 90:145; Hanks 86:83, 85;

Hubon 02:6; Merrill 14:656, 658; Tucker 25:335, 336; Tucker and Reed 39:17-18, pl.1)."

High PeakMine. "Located March 1880. Unpatented. Adjoins Eagle claim which lies to southwest. Formerly included High Peak South and Rossland claims reported by local resident (1957) to have been adjudicated out as mining claims. Developed by a 462-ft partly stoped drift adit drive N. 65 degrees W. along the mineralized shear zone. Adit intersected north-northeast trending cross cut adit driven from Eagle property. Vertical winze was sunk at end of Eagle crosscut on quartz vein 25 ft. northeast of main mineralized zone. A drift on this vein extends southeast from the winze for 60 ft. In 1957, the winze was caved 50 ft. below its collar. Additional caved workings lie to the southeast of the portal of the main adit along the main mineralized zone. The mine was worked sporadically from 1870 to 1902 and from 1928 to 1932. Value of total production estimated by Donnelly (1934, p. 352) to be between $25,000 and $50,000. (Crawford 94:241; 96:339; Donnelly 34:350, 352, pl.4; Hanks 86:83, 85; Hubon 02:7; Merrill 14:656, 659; Preston 90;542; Tucker 25:339-340; Tucker and Reed 39:14, 21, pl.1)."

62.1 miles Back to Main Street, make a left turn, continue east on 78 toward Banner Grade.

64.2 miles Starting down Banner Grade.

64.6 miles **OPTIONAL STOP 4**: Mixed rock sequence exposed in the rock cut on the right side of Banner Grade.

Composition is granodiorite with lineations subparallel to the Elsinore fault zone. Shear zones are evident.

64.4 miles **STOP 5**: Elsinore Fault is on the left. Notice offset drainages, vegetation changes, benches and triangular facets. Approximately 4 miles of the Elsinore Fault trace is visible.

66.4 miles Road to the right goes to the Warlock Mine. Do not take it. The Warlock Mine crosscut-tunnel extends over 1400 feet southward, penetrating Julian Schist, a strand, or branch of the Elsinore Fault, and a host of subparallel gold-bearing quartz veins. The primary targets in the mine were the Shamrock, Neptune and Padlock Veins (approximately 1300 feet S). Over 2 ounces per ton has been assayed at surface outcrops of these veins. Abundant wall rock alteration of meta-sediments can be observed on these vein systems.

Continue down Highway 78 toward Banner.

66.5 miles **OPTIONAL STOP 6**: Gold Panning Demonstration

A tributary drainage fed by erosion of the Warlock mine group veins has contributed fine-grained placer gold to the deeper, coarse-gravels of the stream bed. A few shorels full of these gravels can be hand-washed in a gold pan to prospect for gold upstream. This technique is still used by modern geologists in narrowing target-areas for further exploration. More commonly, approximately 80 mesh, field-screening of dry stream sediments is used by exploration geologists.

68.2 miles Right turn onto dirt road to Chariot Mine. This road roughly follows Chariot Canyon Fault.

70.8 miles **STOP 7**: Chariot Gold Mine entrance on right.

Important Note: Absolutely No Alcoholic Beverages maybe consumed on-site as per the owners request and MSHA Mining Regulations.

The Golden Chariot Mine is perhaps the single most productive mine in the Julian-Banner districts. It is credited with some $800,000 worth of gold production, but a substantial portion of the gold produced from this mine was probably not recorded. This mine was discovered in February 1871, one year after the discovery of gold in the town of Julian. The mine operated from 1871 to 1876 when it was closed upon the death of the owner, Mr. Mark McDonald. The mine laid idle until 1913. Some work was done but it was not until 1923 that the mine was reopened on a larger scale. Work continued until 1931 when mining again ceased. The mine has not been operated on a large scale since 1931. The current owner acquired the property in 1981 and has reconditioned the #1 Working Shaft, erected a new head frame and hoist house, and has constructed a modern mill, assay lab and office.

The ore body of the Golden Chariot mine appears to be located wholly within the Chariot Canyon Fault. The Chariot Canyon Fault is not a single fault, but a broad shear zone at least 600 feet wide, perhaps more. The fault may be recognized beginning at the head of the canyon as crush zones, slickensided surfaces and faulted pegmatite dikes in the road cuts. The Chariot Canyon Fault zone is somewhat unique in that a younger cataclastic (brittle) style of deformation is overprinted (superimposed) upon a much older fault zone with a mylonitic (ductile) fabric. To the west of the mine, the Julian Schist is a Late Paleozoic-Early Mesozoic clastic wedge of marine sediments that was metamorphosed and deformed by Late Cretaceous time. To the east, the granitoids range in composition from tonalite to granodiorite and are part of the enormous La Posta pluton, an intrusive body at least 45 km wide and some 80 km long.

The Chariot Canyon fault separates these two units but affects both. Both the schistose and granitic rocks have a strong mylonitic fabric imposed on them that disappears laterally from the shear zone. The gold is produced from at least four quartz veins, but recent evidence suggest there may be as many as eight. The veins may be up to 10 feet thick. The gangue minerals include pyrrhotite, arsenopyrite, sericite tourmaline, and garnet. Tellurides have been reported from this mine. Recent work indicates that the gold-bearing quartz veins were emplaced about 91 m.y.a. at a temperature of 530 degrees centigrade. The ore forming fluids were probably derived from water that was "sweated" out of micas in the Julian Schist as the huge La Posta Pluton was intruded. The ore forming fluids circulated through dilatant zones in the ancient Chariot Canyon Fault, depositing the gold- quartz veins.

REFERENCES

1 Ellsberg, Helen, "Mines of Julian," La Siesta Press, 1972.

2 "Geology and Mineral Resources of Santa Ysabel Quadrangle, San Diego County, California," California Division of Mines and Geology Bulletin 177, 1958.

3 Kofron, Ronald J., "Age and Origin of Gold Mineralization in the Southern Portion of the Julian Mining District, Southern California," Unpublished Master's Thesis, San Diego State University Department of Geological Sciences, 1984.

4 Koschmann, A. H. and Bergendahl, M. H., "Principal Gold-Producing Districts of the United States," U.S. Geological Survey Professional Paper 610, p. 76, 1968.

5 Pourade, R. F., "The Glory Years," Volume 4, The History of San Diego, Union-Tribune Publishing Co., San Diego, CA, p. 274, 1964.

6 Weber, F. Harold, Jr., "Geology and Mineral Resources of San Diego County, California," California Division of Mines and Geology County Report 3, 1963.

Life Beyond Gold:
A New Look at the History of Julian, California

KATHRYN A. JORDAN

IN THE MID-NINETEENTH CENTURY, MINING TOWNS appeared throughout California only to become ghost towns as miners hurried to new deposits elsewhere, leaving abandoned buildings in their wake. The town of Julian, founded during a gold strike in 1869, did not disappear but, instead, became a community. Men and women chose to stay in Julian because of the close relationships that developed among families and the viability of agriculture, particularly apples, in the region. This article describes the founding of Julian and examines the factors that allowed the town to survive beyond the Gold Rush years.[1]

According to local folklore, Native Americans living in the Cuyamaca Mountains knew about gold deposits in the surrounding hills but kept it a secret until they could use it to their advantage. Eventually, they began trading gold dust for trinkets from padres at Santa Ysabel Mission and, later, showed them where gold could be mined.[2] A more credible tradition suggests that A. E. (Fred) Coleman, a former slave, first discovered gold near present day Julian in January 1869. Coleman, who previously mined in northern California, lived with an Indian family near Wynola, about four miles southwest of the Santa Ysabel ranch. While watering his horse in a creek, he noticed a reflective material in the creek bed. He panned through water, stone and soil, and exposed a few small nuggets of gold.

He quickly formed the Coleman mining district and nearby Emily City mining town. Word quickly spread to the surrounding hills and San Diego.[3]

Kathryn Jordan, *a candidate for the Master of Arts degree in history at the University of San Diego, plans to compare the histories of Julian, Big Bear and Calico to show why Julian and Big Bear survived after the mining declined. She thanks Dr. Iris Engstrand, the San Diego Historical Society, and the San Diego County Library, Julian branch, for their assistance.*

Later that year, four cousins, Drury and James Bailey and Mike and Webb Julian, arrived in Temecula from their home state of Georgia. On their way from Nevada to the Southern California coast, the men learned of proposed plans for a Pacific Railroad line from Yuma, Arizona, to San Diego. Short on supplies, they sent James to San Diego to obtain both rations and information about the railroad. On his return, James met Mr. Harrall, a resident of the Cuyamacas, who was headed to San Diego to trade bacon for supplies. Harrall told him that the land in the Cuyamacas had abundant wild hogs and unpanned streams. The cousins figured that there would not be work in San Diego until the railroad was completed so they headed back to Arizona. While crossing the Cuyamacas in November 1869, Drury Bailey decided to homestead there. The others also stayed and, the following spring, planted barley on land that would become the town of Julian.[4]

In February 1870 two separate gold deposits were discovered near the Baileys' camp. Drury Bailey uncovered a quartz ledge, "Warrior's Rest," a few days before H. C. Bickers discovered what would become the Washington Mine.[5] Together the men formed the Julian Mining District on February 15, 1870, in order to protect the interests of the growing number of miners in the area. The district appointed Mike Julian as its first district recorder. Around this time, Drury Bailey titled the town "Julian" after his cousin, Mike, because he was the most handsome man in town, and it sounded better than "Bailey" for a town name. By March, twenty claims had been filed and the *San Diego Union* began running reports of gold, encouraging the migration of prospectors to Julian and the surrounding Cuyamaca towns like Branson City, Eastwood, and Coleman City (formerly Emily City).[6]

Quickly, landowners moved to try to control the goldmines. Robert Allison, John Treat, Juan Manual Luco and Isaac Hartman had bought land south of Julian from Don Agustin Olvera in 1869, intending to use it for lumber and grazing.[7] News of gold caused them to reevaluate their plan. Two months after the formation of the Julian Mining District, Luco and the other men began legal proceedings to alter the boundaries of the Cuyamaca Grant in order to include the gold fields within their property lines. Sherman Day, U.S. Surveyor-General for California, directed Deputy Surveyor James Pascoe to conduct a government survey but it soon became apparent that Luco and Treat influenced Pascoe to alter the grant boundaries. On May 25, 1870, the grant owners met with the Julian miners and offered a compromise: mining could continue so long as royalties were paid to the grant owners for any gold extracted. The miners quickly rejected this idea, knowing some mines would be taxed up to fifty percent of their earnings.[8] Within that week, Julian and San Diego residents met at Horton Hall in San Diego to discuss action against the grant owners.

Julian residents initiated what would become a long legal battle for access to the goldmines. They formed a committee known as the Defense League and began fundraising for the payment of legal fees to attorney George Yale. Soon afterward, Judge Benjamin I. Hayes began working independently on the miners' behalf. By January 1, 1871, Hayes assumed full legal responsibility for the case from Yale. Throughout the summer of 1870 he conducted a field survey of the area, visited and interviewed Indians and settlers in the region, and documented his findings. Around the same time, the Defense League hired engineer Charles J. Fox to conduct a survey of the Cuyamaca Grant. Fox discovered, as expected, the grant boundaries ended roughly six miles south of the Julian mines. In order to pay for legal fees accumulated by the efforts, miners contributed a dollar each month and held fundraisers like dances and cake sales for the cause. The entire town celebrated when the new U.S. Surveyor General for California, J. R. Hardenbergh, closed the case on April 5, 1871, without changing the boundaries of the grant. However, the grant owners managed to have the case reopened in November 1872 on a technicality. In the end, however, the miners prevailed and the matter was closed indefinitely on April 25, 1873. President Ulysses S. Grant approved the official land grant boundaries on December 19, 1874.[9]

Julian survived beyond the mining boom due to the large number of women and families, not just single men, who migrated to the region. One local historian suggested that Julian did not became a "ghost town" like other California mining camps largely because "the women of the families, and the men not actively engaged in mining, had found that the soil was miraculously productive. Many of the families chose to stay in the district."[10] Several women in town managed their own hotels or boarding houses for miners. Jennie Garrett Lane, who was known as "Grandma Lane" throughout the town, ran a boarding house and worked as a midwife. Margaret Tull Robinson ran the Hotel Robinson with her husband, Albert. After Albert's death, Margaret continued to run the hotel herself until the Jacobs family purchased it around 1919.[11] As a teenager Helen Jacobs helped do "everything" around the hotel. She waited tables, cooked in the bakery and café, and washed dishes. When the hotel was busy, Annie Grosskopf, a Canadian Indian known in town as Cris Cross Annie, also helped do odd jobs around the hotel.

Children and their teachers gather in front of the Julian schoolhouse, ca. 1890. ©SDHS #80:7545.

While men worked predominantly in the mines, women kept stores and hotels open for business.[12]

A school for the children lent an aura of permanence to the town. James Jasper noted "people always flock to a new gold camp; not men alone, but women as well, and where there are women there are children; schools must be provided, and you can trust the miners for that."[13] Julian's first school opened its doors in 1870; a school in neighboring Banner followed in 1872. At the time, students used outhouses and a wood-burning stove. In a letter from Julian to northern California, one new resident noted "the place is assuming an apparent permanence, since the discovery of the mines. There is a large and commodious school-house and an organized school district, numbering over 100 children of school age, with school in operation, and every Sunday, as Mrs. Partington says, we hear the 'gospel dispensed with.'"[14] The first school year began May 15, 1870, and closed June 30, 1870. For seven weeks of teaching, the teacher earned $450, nearly as much as she would make in the following school year. An average of twenty-seven students attended in this month and a half.

Many men and women met their future partners in Julian. Teachers, often young, unmarried women, were sought after by single miners. The first Banner teacher, Edith Shaw, married Isaac Ijams shortly after her tenure. Her successor Elizabeth Kelly married Chester Gunn. Susan Stormes, Julian's first teacher, married Charles Leonard Evans in 1874. According to Jasper, "the last wedding

America Newton, a former slave from Independence, Missouri, owned an 80-acre homestead near Julian and provided laundry services to the town. She is pictured here, ca. 1910. ©SDHS #11109.

caused the trustees of both districts to declare a boycott of school marms," but after the bachelor miners threatened the trustees, "the trustees realizing their danger, revoked the boycott and a feast and dancing followed putting an end to the rebellion."[15] Many of the couples who met in Julian remained and raised their families in the area.

Julian's melting pot of ethnicities contributed to a vital business community that included a cabinetmaker, painter, farmer, hatter, druggist, brick mason, cooper, blacksmith, and baker. Census records indicate that English, Polish, Welsh, Jewish, Italian, and African American residents called Julian home in the town's earliest days. "Count" Dwarskowski, who claimed descent from Polish nobility, owned one of the town's general stores. Joseph Swycaffer ran the town's only butcher shop.[16] America Newton, a former slave from Independence, Missouri, was also well known around town. She made her living washing clothes, and she was often seen riding in her buggy with clean laundry to be delivered. Albert and Margaret Robinson, an African American couple who met in Julian, owned and operated the Hotel Robinson. Employees remembered Mrs. Robinson as "very prim and energetic" but "quiet-spoken." According to one local, they were considered to be "fine colored people."[17]

The Wilcox store in Julian served as a general store, post office, and stage coach stop, ca. 1885. ©SDHS #3593-G.

Native Americans and Chinese did not fare as well as other ethnic groups in Julian. Indians sold fruits and vegetables to miners and worked as laborers on road building projects. White residents could be intolerant of their behavior, particularly when they violated social norms. In August 1890 James Jasper wrote the Sentinel that "the mob of drunken Indians of last Sunday…was a disgrace to the town."[18] There were few Chinese residents due to prejudice against them. The 1870 census reported only one "domestic servant" of Asian descent. Some restaurants and hotels, even twenty years after the discovery of gold, advertised that they did not employ Chinese cooks. The Chinese who lived in Julian in the late nineteenth century eventually moved to more tolerant communities in San Diego and Los Angeles.[19]

Julian hosted a small but stable number of downtown businesses. A San Diego Union article noted "Julian is what may be called a 'rising town,' not only with reference to its altitude, but as to its rapid settlement and building up."[20] Six hundred people populated the town soon after the discovery of gold and more arrived over the next few years. Between 1870 and 1880, Julian boasted two hotels, five stores, two cafes, two blacksmith shops, two livery stables and several saloons. The town did not, however, have a bank until September 1870. In their absence, general stores served as banks. Markets in San Diego and Los Angeles gave credit to the stores in Julian for the gold deliveries they received and the stores, in turn, gave credit to the miners to use for groceries or, the most widely purchased good, alcohol. Historian Helen Ellsberg noted, "Aside from the difficulty of bringing in such store-bought supplies as coffee, sugar, and flour, living was easy in the mountains."[21]

Boarding houses and hotels catered to Julian's growing population. Before that, people slept in tents. Emily City, for example, was considered a "tent city" prior to the creation of boarding houses in the area. The Pioneer Hotel, run by Mrs. M. Clough, "could accommodate forty guests, who, the year before, would have been sleeping outdoors."[22] George and Katherine Hoskings built and ran the Julian Hotel (not to be confused with the Hotel Robinson). After George's death in 1874,

The Julian Hotel, built in 1872, was operated by George and Katherine Hoskings. The hotel burned down around 1900 and its name was later adopted by the owners of the Hotel Robinson. ©SDHS #3592-G.

Katherine remarried and continued running the hotel until her death in 1885. As a young teenager, Suzie Taylor McPherson waited tables for Mrs. Hoskings at her hotel. By the age of sixteen, she had taken management of the Pioneer Hotel, renamed it the Mountain View Hotel, and ran it for George Keener, the property owner.[23]

Restaurants served as working kitchens for everyday meals as well as places of entertainment. The Hotel Robinson had a number of "regulars" who relied on Mrs. Margaret Robinson's cooking for their sustenance. Joe Marks, Dr. Hildreth, the stagecoach driver, a salesman named Drummers, and a number of teachers visited the restaurant on almost a daily basis. Similarly, "Mother Lane" ran a miners' boarding house where "she…cooked for more miners and fed more hungry people free than any woman in California."[24] While she may not have profited from her generosity, word of her cooking spread and contributed to her reputation for good, hearty meals. Police Captain Francis Marion Hopkins served dual roles as saloon owner and policeman. He even held court in his bar and managed to "dispense justice and whiskey in the same room without mixing results." A newspaper correspondent writing under a fictitious name, Nil Desperandum, described Julian in this way: "Though the town is not very extensive, when we come to think that it has been settled but little over two years, its size and the completeness of its amusement resources, is surprising."[25]

Transportation to and from San Diego also helped business in the mountain town. Joseph Yancey originally moved his family to Julian from Georgia in 1870, not for gold, but to raise horses. He used his horses to transport ore between the mines and stamp mills, and merchandise between Julian and San Diego. Before consistent mail delivery, Chester Gunn operated a pony express mail service from Julian to San Diego beginning in April 1871. Because he traveled alone, he could take trails and short cuts that wagons could not, thus cutting down delivery time. Gunn charged "ten cents a letter and carried letters and small packages," and received a fair amount of business because people wanted reliable mail service, even for a small fee. Demand for a pony express service eventually disappeared as stages began to make the trek from San Diego to Julian in a single day. Regardless of the method, speed of delivery improved, thus benefiting both personal and business communications.[26]

Eventually, news of new gold discoveries in Arizona and Nevada drew miners away from Julian and the population dropped to around one hundred by the mid- to late 1870s. Nevertheless, the town survived. In 1881, the Julian and Banner mining districts combined to form the new Julian district. The discovery of the Gold King and Gold Queen mines in 1888 gave new life to

James Madison's orchard and home, 1890. Madison and Thomas Brady are credited with bringing the first apple seedlings to Julian in the 1870s. ©SDHS #3593.

Certificate awarding W. L. Detrick of Julian a gold medal for his apples at the Louisiana Purchase Exposition in St. Louis, 1904. ©SDHS #91:18564–1230.

mining and a boost to the town's businesses.[27]

Julian's continued prosperity had a good deal to do with its apple, and later pear, orchards. James Madison migrated to San Diego from the east coast in 1867 looking for land suitable for raising horses. He recognized the soil and climate would be suitable for growing apples. Madison and Thomas Brady planted the first apple trees in Julian in the 1870s. The *San Diego Union* reported the news of Madison's apple harvest on November 11, 1880. One former resident remembered, "everybody went to agriculture" after mining production halted in the late 1890s. The Baileys, Wellingtons, and Horace Wilcox all planted large orchards around that time. Productive harvests helped Julian apples win eight gold medals in 1907 at the Jamestown Virginia Exposition.[28]

Close-knit friendships and an active social life kept people from leaving Julian. With what appears to have been a light-hearted, pioneer spirit, residents seemed to genuinely enjoy one another's company. According to one early historian, "Julian was never the 'hell roarin' town that is commonly associated with mining camps." Horace Fenton Wilcox remembered "during the gold rush Julian and Banner was pretty tough places, but I reckon they wasn't any tougher'n most minin' camps of that time. Every other place of business was a saloon, a gamblin' joint, or a dance hall; but on the whole things was pretty orderly...respectable women was perfectly safe."[29] One visitor from San Diego returned to the city and noted, "the mountain people understand life. When they have a 'time,' they have a good time."[30]

The Julian Chamber of Commerce celebrated the continuing prosperity of the town by sponsoring a float in the Julian Apple Days Parade, October 4, 1959. ©SDHS UT #85:a5063, *Union-Tribune* Collection.

Practical jokes occurred with regularity. Tom Daley, the town butcher, allowed a German couple, Waldemar and Maggie Wilson, to live in the back of his shop. Waldemar mined for a living and Maggie raised and sold chickens in town. Daley thought it funny to take Maggie's chickens without her knowledge in an attempt to fluster her. Before Daley could have a good laugh, he noticed the eggs from his own chicken coops were gone. Maggie had gotten the best of him. Drury Bailey, notorious for his sense of humor, left a town dance early one evening and snuck back into the room where all the babies belonging to party goers slept. Thinking it funny to move the children from crib to crib, he switched them around and hopped on the last stage out of town to the county seat in San Diego. When the dance ended, the mothers picked up their babies and headed home, not knowing that they had taken another woman's child. It was not until Drury returned three days later to explain his mischief that they mothers were able to sort out the confusion. Engaged couples like Lucy Wilcox and George Dannals even tried to keep their romance a secret because they did not want to be victims of a prank.[31]

The most popular and frequent social gatherings in Julian revolved around dancing. The Wilcox family usually provided the music. Horace Fenton Wilcox played the bass viol, his father the fiddle, and the Wilcox daughters played the piano. Mr. Wilcox used this as a supplemental source of income when his mine was not producing. Some dances were thrown for the simple purpose of dancing. Other times, dances were held as fundraisers for a new town hall or road construction. James Jasper recalled, "In pioneer days everyone danced in Julian… there were more than the usual number of matrons with young babies. The dance hall was fully equipped with a ladies dressing room in which the babies were put to bed, while the mothers 'tripped the light fantastic toe.'"[32] Often, the *Julian Sentinel* announced the details of an upcoming dance, and followed up with a report on the dance's success in fundraising along with details about what was served for dinner. The St. Patrick's Day dance in 1891, for instance, began around 9:30 p.m., and continued until 12:00 midnight, at which time the thirty couples present ate supper at Mrs. Williamson's restaurant. Following their fest, "dancing was again resumed and continued till 4:30 in the morning." The dance, held as a fundraiser for a new floor, raised about $15.[33]

Church attendance provided another form of fellowship among townspeople.

The first group of miners to arrive in Julian included three Baptist ministers. Rev. Thomas Jackson Wood, a missionary from England, began services at the Julian Community Baptist Church in 1885. Services were originally held in the one- room Julian schoolhouse until Drury Bailey donated land to be

Alice Genevieve Barnes (Mrs. Franklin), pictured by an apple tree in April 15, 1959, grew up in early twentieth- century Julian. Her reminiscences about everyday life, social relationships, and local businesses, made a significant contribution to the town's historical record. ©SDHS UT #85:a1838–4, *Union-Tribune* Collection.

used for a new church building. The new building held its first church service in December 1891 and, shortly thereafter, Rev. Frank L. Blanc was hired as the church's full-time pastor. The *Sentinel* advertised for volunteers to assist in the organization of a church Sunday school and called on families to contribute to the effort. Aside from Sunday morning services, "song services" were held in the evenings to give "the young couples a chance to get together."[34]

Small town gatherings continued into the early twentieth century, long after the gold mines had ceased production. The first Apple Day celebration was held in 1909 to celebrate the year's harvest and a new stage in the life of Julian. The day was overcome with windy conditions so men and boys held the tents in place for food serving while "the women and girls had to tote the food across the street to the Julian Hotel." Julian and other rural towns like Escondido, Ramona and Poway, formed a town baseball league, which provided recreation for the men in town. Fourth of July celebrations were elaborate affairs. People from miles around gathered in Julian for a barbeque, horse racing, relay races, and dancing.[35]

Today Julian is famous for apples and apple pie. Orchards produce several varieties, including Pippins, Arkansas Black, Lady, and Granny Smith. Tourists and locals enjoy Julian's apples in pies, ciders, and jellies available throughout the county.[36] Many of the town's original buildings remain standing and are used for shops, restaurants, and other small businesses. While most California mining towns gained fame for their boom years, Julian boasts of a history beyond the Gold Rush.

NOTES

1 This paper draws on sources held at the San Diego County Public Library, Julian branch, and San Diego Historical Society. These archives provide adequate but limited evidence about life in early Julian. The Julian Historical Society, however, promises to make available previously unread letters, diaries, and other materials related to the history of the town. The Society's archives, in storage for a number of years, are presently in the process of being archived, indexed, and preserved, in order to make them available to the public. They should be available at the Society's new home, the old Witch Creek School, most recently used as the town library, in the not-too-distant future.

2 Dan Forest Taylor, "Julian Gold," *Federal Writer's Project*, February 8, 1939, 2. Mission Santa Ysabel was founded in 1818 as a branch, or "asistencia," of the Mission San Diego. Thanks to Dr. Iris Engstrand for this information.

3 Richard F. Pourade, *The Glory Years*, (San Diego: Union-Tribune Publishing Comopany, 1964), 53–54; Helen Ellsberg, *Mines of Julian*, (Glendale, CA: La Siesta Press, 1972), 13; Gale W. Sheldon, "Julian Gold Mining Days," masters thesis, San Diego State College, 1959, 17. At the time of Coleman's discovery, three known families lived in the area: the Harrall, Brady, and Webb families. Horace Fenton Wilcox, "How Julian Mines Were Discovered," [n.d.], San Diego Historical Society Archives (hereafter SDHS), Subject File "Julian, CA," 2. There is little, if any, additional

information about Emily City, other than the fact that the "city" consisted primarily of tents.

4 Myrtle Botts, *History of Julian* (Julian, CA: Julian Historical Society, 1969), 1–5. The Baileys and Julians made an agreement to meet in New York City on a given day in 1867 and travel together to San Francisco by way of Panama. Frank Bailey, Drury and James Bailey's brother, also traveled with them from New York to the West Coast. At some point in the travels around Nevada, Frank separated from the group but rejoined them in Julian in 1870. Drury ("Drue") Bailey failed to arrive in New York by the chosen day, and as a result traveled west independently through Missouri. He stopped to prospect in Montana, Idaho, Utah and Arizona before arriving in Nevada where, by chance, he was reunited with his family.

5 Charles LeMenager, *Julian City and Cuyamaca Country*, (Ramona, CA: Eagle Peak Publishing Company, 1992), 42. The Baileys set claim to their gold deposit, "Warrior's Rest," a few days earlier than Bickers's discovery on February 20, 1870. Considered a 'pocket claim,' Warrior's Rest was deserted, whereas the Washington Mine continued production for years. Taylor, "Julian Gold," 4–7.

6 Botts, *History of Julian,* 5; Ellsberg, *Mines of Julian*, 17.

7 Robert Allison later founded La Mesa, a suburb of San Diego. Allison Avenue was named in his honor. See Liz Neely, "On Allison, Council Says Its Aim is True," *San Diego Union-Tribune,* April 5, 2007.

8 LeMenager, 59–62.

9 Ibid., 62–66, 72–75. Hayes, Samuel Ames, former resident of the Cuyamacas, and their Indian guide, Chono, set out in July 1870; James A. Jasper, "Trail-Breakers and History-Makers," [n.d.], SDHS, 22.

10 Ray Redding, interview by Jeanette Washburn, *KGB Radio* transcript, October 19, [n.y.], SDHS Subject Files "Julian, CA."

11 Sources are unclear as to the exact year that the Margaret Tull Robinson sold the Hotel Robinson to Martin Jacobs. They also are unclear about when the Jacobs family changed the hotel name to the Julian Hotel. Some sources, including Alice Barnes, recall that Martin Jacobs bought the hotel in 1919. See Barnes, "Alice Genevieve Barnes: Gold Mines and Apple Pie," 70. Others believe it occurred earlier, in 1918. See LeMenager, *Julian City and Cuyamaca Country,* 149, 190. The Julian Gold Rush Hotel, which now occupies the original Hotel Robinson, states the transfer of ownership happened in 1921, and that the name change from Hotel Robinson to Julian Hotel took place sometime during the forty- seven years the Jacobs family ran the hotel. See http://www.julianhotel.com/history.html (accessed March 5, 2008).

12 Alice Genevieve Barnes, interviewed by Scott Theodore Barnes, n. d., San Diego County Library, Julian Branch. Barnes published a limited edition of these transcripts: Barnes, "Alice Genevieve Barnes: Gold Mines and Apple Pie," (n.p: private printing, 1999), 18–19, 91.

13 James A. Jasper, "Julian and Round-About," [1928], 7–8, SDHS.

14 "Letter from San Diego," *Sacramento Daily Union*, November 16, 1870. Local Indian children did not attend school in Julian, but were certainly visible and active in the town. The Indian children typically attended an Indian boarding school on the reservation in Volcan. See Barnes, "Alice Genevieve Barnes: Gold Mines and Apple Pie," 73.

15 Jasper, "Trail-Breakers," 66.

16 Jo Ann Cornelius, ed., *1870 Census of San Diego County California* (San Diego Genealogical Society, 2nd ed., 1995), 99; Taylor, "Julian Gold," 11; Jasper, "Trail-Breakers," 10; Martin Jacobs ran a butcher shop from 1915–1920. At the time the Jacobs's butcher shop was the only place in Julian to get ice cream. See Barnes, "Alice Genevieve Barnes: Gold Mines and Apple Pie," 91, 98.

17 Botts, *History of Julian,* 24–25; Jimmy Thornton, "Native Remembers a Different Julian," *San Diego Tribune*, January 4, 1988, SDHS Subject Files "Julian, CA." One author described them as "hard working, thrifty and resourceful, traits which helped them prosper and break down barriers of discrimination." Paula Parker, "Heyday of Julian Hotel," *Los Angeles Times*, February 3, 1980, II:CC:1, SDHS Subject Files "Julian, CA."

18 Ellsberg, *Mines of Julian*, 65; James A. Jasper, ed., *Julian Sentinel*, August 29, 1890. After a man was killed in front of the newspaper office and the murder failed to make the headlines, James A. Jasper decided a change was necessary. He purchased the newspaper from A. J. Jenkins in 1887 and began "telling the news as it happened." See Ellsberg, *Mines of Julian*, 25; LeMenager, *Julian City and Cuyamaca Country*, 124.

19 Cornelius, *1870 Census of San Diego County California*, 100; *Julian Sentinel*, May 30, 1890, SDHS Subject Files "Julian, CA"; Ellsberg, 67.

20 Leland Fetzer, *Good Camp: Gold Mines of Julian and the Cuyamacas* (San Diego: Sunbelt Publications, 2002), 16.

21 Ellsberg, *Mines of Julian*, 17, 22; Taylor, "Julian Gold," 11; *San Diego Union*, September 1, 1870, col. 2, 4.

22 Cited in Sheldon, "Julian Gold Mining Days," 68.

23 Jasper, "Trail-Breakers," 109; Suzie Taylor McPherson, interviewed by Edgar F. Hastings, April 1, 1959, transcript, SDHS.

24 Barnes, "Alice Genevieve Barnes: Gold Mines and Apple Pie," 98; Jasper, "Trail-Breakers," 148.

25 Ellsberg, *Mines of Julian*, 23; Sheldon, "Julian Gold Mining Days," 74.

26 Virginia Bruett Wyatt, *The History of Camp Virginia* (n.d., private printing, 1983), San Diego County Library, Julian Branch; LeMenager, *Julian City and Cuyamaca Country*, 94, 96.

27 Ellsberg, *Mines of Julian*, 25.

28 LeMenager, *Julian City and Cuyamaca Country*, 165; Rufus Collor, interview by Robert Wright, transcript, December 29, 1974, SDHS; "A Short History of Julian, California," *KGB Radio* transcript, San Diego, February 10, 1963, SDHS Subject Files "Julian, CA."

29 Taylor, "Julian Gold," 33; Horace Fenton Wilcox, "Memories of the Gold Stampede to Julian," *Touring Topics*, February 1932, 17. The interesting thing about Wilcox's recollection, is it is one of the few, albeit brief, mentions of gambling, dance halls, or the notion of non-respectable women in Julian. On the contrary, Dan Forest Taylor goes so far as to say gambling halls and dance hall girls were intolerable in Julian, especially compared to other gold mining regions like the Yukon and Klondike. See Taylor, 34. Helen Ellsberg alludes to the notion that Jasper's manuscript has been tampered with over the years, by families who did not want their personal affairs aired to the public. As a result, "all reference to shady ladies and a red light district [are] missing." Ellsberg, *Mines of Julian*, 25.

30 Sheldon, "Julian Gold Mining Days," 151.

31 Herbert Lockwood, "The Skeleton's Closet," [publication not noted], July 7, 1968, SDHS Subject Files "Julian, CA"; Botts, *History of Julian*, 22; Winifred Davidson, "Tales of the Old Southwest," *San Diego Union*, September 3, 1939.

32 Wilcox, "Memories," 17; Suzie Taylor McPherson, interviewed by Edgar F. Hastings, April 1, 1959, transcript, SDHS.7; Jasper, "Trail-Breakers," 29; *Julian Sentinel*, March 20, 1891, SDHS Subject Files "Julian, CA."

33 Residents also rallied around student performances held by the Julian school. Talent show performances included a number of songs, recitations, and readings. The money earned from the event was "used in procuring portraits of Washington, Grant, Longfellow and Beethoven for the new school house." *Julian Sentinel*, August 8, 1890, SDHS Subject Files "Julian, CA."

34 Taylor, "Julian Gold," 4; LeMenager, *Julian City and Cuyamaca Country*, 148, 150; *Julian Sentinel*, July 11, 1890; Donald A. Moore, interview by San Diego Historical Society, ND, SDHS.

35 "A Short History of Julian, California," *KGB Radio* transcript, San Diego, February 10, 1963, SDHS Subject Files "Julian, CA"; LeMenager, *Julian City and Cuyamaca Country*, 143; Charles Kelly, interviewed by Edgar R. Hastings, transcript, July 15, 1957, SDHS.

36 Ernie Grimm, "Julian's Apple Survivors," *San Diego Reader*, November 8, 2001.

Welcome to the California Geological Survey

IN APRIL 1860 THE CALIFORNIA LEGISLATURE ESTABlished the Geological Survey of California, which has evolved during its long history of service, and several name changes, into today's modern California Geological Survey (CGS).

CGS is dedicated to the fulfillment of its mission to provide scientific products and services about the state's geology, seismology and mineral resources that affect the health, safety, and business interests of the people of California.

CGS geoscientists, engineers and support staff take great pride in upholding the heritage of an organization that has a 150-year legacy of providing the highest quality of scientific and engineering products and services to the State of California.

—"ALTIORA PETIMUS" (We Reach Higher)

Reference: http://www.conservation.ca.gov/cgs/Pages/Index.aspx

The following CGS notes (Note 12 and Note 24) contain general information relating to mining that might answer common questions.

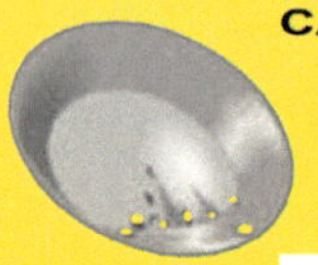

CALIFORNIA GEOLOGICAL SURVEY

GOLD

NOTE 12

CALIFORNIA STATE MINERAL

The accidental discovery of gold in 1848 at Sutter's Mill in Coloma started a bonanza that brought California fame and gave it the title of the "Golden State." The Gold Rush of 1849 and the subsequent influx of settlers led to California becoming the 31st state in 1850.

NATURAL OCCURRENCES

The concentration of gold ore, and the market value of gold determine whether a deposit is a mineable orebody. The highest grade deposits are associated with quartz veins. Gold also occurs as disseminated particles incorporated during magmatic rock formation or during subsequent chemical alteration of the host rock. Primary gold occurrences are termed lode deposits. Mineralized rock and gold-bearing veins release gold particles during the weathering process. Because of its high specific gravity and resistance to weathering, these sedimentary gold particles are easily concentrated by streams and rivers to form placer gold deposits.

MINING METHODS

There are three types of gold mining today: underground mining of high-grade lode and placer deposits, dredging of surface placer deposits, and open-pit mining. A technique called heap leaching is commonly used to remove finely disseminated gold from low-grade ore. In this process, mounds of crushed ore are placed on an impermeable pad and sprayed with a dilute cyanide solution. The cyanide solution percolates through the ore and dissolves fine gold particles. The gold is then electrolytically recovered from solution and poured into ingots. Gold is also a byproduct of sand and gravel production and base metal (copper, lead and zinc) mining.

GOLD DEPOSITS IN CALIFORNIA

California's most important gold deposits have been found in the Sierra Nevada, Klamath Mountains and Mojave Desert. Significant deposits have also been developed in the Peninsular and Transverse Ranges and the northern Great Valley. Unmined low-grade deposits occur statewide. In the Coast Ranges, low-grade gold deposits are associated with low-temperature mercury mineralization.

GOLD PRODUCTION IN CALIFORNIA-2001

In 2001, California ranked fourth in the United States in gold production. Approximately 449,200 troy ounces were produced—worth about $122.3 million.

PROPERTIES AND USES

Gold is one of the earliest metals known and used by humans. It resists corrosion and chemical interaction. It will not disintegrate when exposed to oxygen, water, salt, or any other naturally-occurring material. Gold's durability accounts for the almost perfect condition of coins and artifacts fashioned from it thousands of years ago.

Gold's most important use is in computers, weaponry and aerospace. It is used where consistent, reliable performance under all conditions is essential. The electronics industry has tried to find substitute metals and alloys, but gold's exceptional resistance to corrosion and tarnish is still unequaled.

CALIFORNIA GEOLOGICAL SURVEY
CALIFORNIA CONSERVATION

Revised 4/2002

HISTORIC CALIFORNIA GOLD MINES

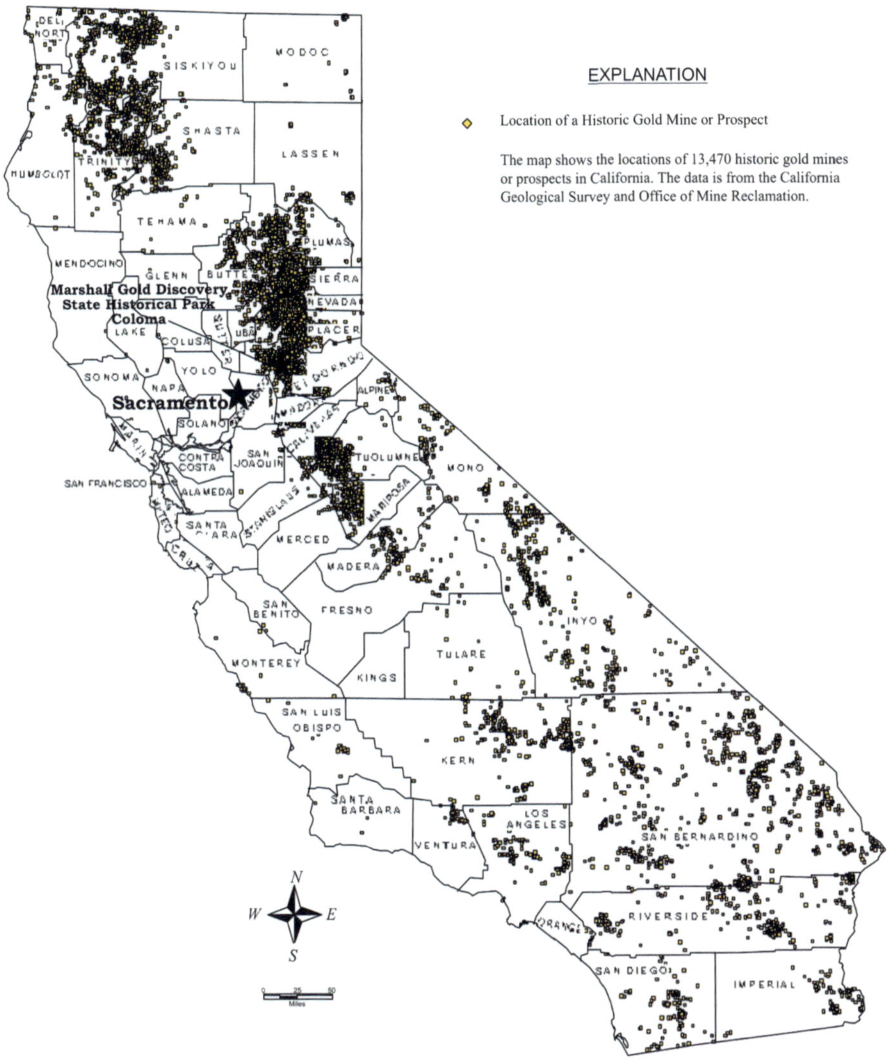

EXPLANATION

◇ Location of a Historic Gold Mine or Prospect

The map shows the locations of 13,470 historic gold mines or prospects in California. The data is from the California Geological Survey and Office of Mine Reclamation.

To learn more about GOLD, visit our California Geological Survey Library in Sacramento.
California Geological Survey Library
14th Floor
801 K Street, MS 14-34
Sacramento, CA 95814-3532
Open Monday-Friday 8am-5pm (closed on state holidays)
(916) 327-1850

Visit the California State Mining and Mineral Museum in Mariposa, along historic Highway 49.
P.O. Box 1192
Mariposa, CA 95338
(209) 742-7625

Two-inch piece of crystalized gold, Mariposa County. *Photo by Jim Spriggs.*

GOLD FACTS

Chemical symbol: Au

Atomic number: 79

Atomic weight: 196.967

Specific gravity: 19.3 (Twice as heavy as lead; 1 cubic foot weighs over half a ton.)

Hardness: 2.5 to 3 on the Mohs hardness scale

Tensile strength: 19,000 pounds per square inch

Melting point: 1,945°F

Boiling point: 5,378°F

- Face centered-cubic crystalline structure.
- Most malleable and ductile of all metals.
- Inert, therefore does not corrode. (Referred to as a noble metal.)
- Good conductor of electricity.
- One ounce of gold can be stretched into a wire more than 40 miles long.
- Gold can be worked into a layer measuring 1 millionth of an inch.
- There are certain produced acid and alkaline compounds that will dissolve gold. The best known is ***aqua regia***, a potent combination of hydrochloric and nitric acids.

Selected Gold-Related Publications of the California Geological Survey

- CD98-001 *California Gold Mines: A Sesquicentennial Photograph Collection on CD-ROM*. 1998
- B193 *Gold Districts of California, Sesquicentennial Edition: California Gold Discovery to Statehood* (includes new mine list). Scale 1:1,000,000. 1998
- MM009 *Map of California Historic Gold Mines. The California Gold Discovery to Statehood Sesquicentennial (1998-2000) Edition*. Scale 1:1,500,000. 1998
- *The Elephant as They Saw It: A Collection of Contemporary Pictures and Statements on Gold Mining in California (1500-1860)*. 1997
- SP098 *Fluvial Geomorphology and River-Gravel Mining: A Guide for Planners, Case Studies Included*. 1990
- SP103 *Mines and Mineral Producers Active in California (1997-98)*. Scale 1:500,000. 1999
- SP034 *Geology of Placer Deposits*. 1970.
- SP041 *Basic Placer Mining*. 1946
- SP087 *Placer Gold Recovery Methods*. 1986

To order one of these publications (or any other California Geological Survey publication) call 916-445-6199.

Map compilation by Les G.Youngs. Graphics by Dinah D. Maldonado, California Geological Survey.

Offices of the California Geological Survey

Publications and Information Office
801 K Street, MS 14-33
Sacramento, CA 95814-3532
(916) 445-5716

Southern California Regional Office
655 South Hope Street, Suite 700
Los Angeles, CA 90017-3231
(213) 239-0878

Bay Area Regional Office
185 Berry Street, Suite 210
San Francisco, CA 94107-1728
(415) 904-7707

www.conservation.ca.gov

THE RESOURCES AGENCY
MARY NICHOLS
SECRETARY FOR RESOURCES

STATE OF CALIFORNIA
GRAY DAVIS
GOVERNOR

DEPARTMENT OF CONSERVATION
DARRYL YOUNG
DIRECTOR

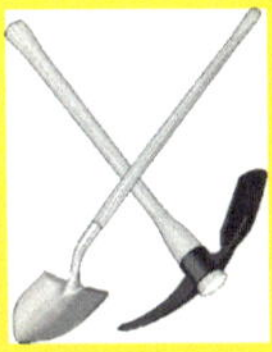

CALIFORNIA GEOLOGICAL SURVEY

HINTS FOR GOLD PROSPECTORS

NOTE 24

The search for gold still offers a special excitement for many people. They hope to "strike it rich" by panning gold. What they do not realize is that most placer deposits have been thoroughly reworked at least twice—first by those who entered immediately after the initial boom periods of the late 19th century to glean the lower grade deposits and tailings and later by prospectors during the Great Depression of the 1930s. During the "rush" of the 1970s and 80s, when the price of gold was at an all-time high, many of these areas were worked once again.

Nevertheless, a chance for success may still remain if you choose favorable areas after a careful study of their geology and mining records. If you are prepared to undergo a certain amount of hardship, possess a vehicle capable of traveling the roughest and steepest roads, and are not discouraged by repeated disappointments, a prospecting trip may prove to be interesting.

The search itself is often its own reward, but there are things you should consider before setting out on a prospecting expedition.

❑ It is a conservative estimate that about 1 in 1,000 people who have prospected in the western United States ever made a "strike." Most of the gold mining districts in the west were located by pioneers, many of whom were experienced gold miners from Alabama and Georgia. Even in those pioneering times however, only a small number of prospectors found valuable deposits. However, the best chances of success are in known productive areas, rather than in unknown areas.

❑ Lode gold is gold within solid rock. The areas that are likely to contain valuable lode deposits of gold have already been explored carefully and thoroughly. These days the inexperienced prospector without ample capital has very little chance of discovering a lode rich enough to develop. Such known areas in California are the Mother Lode, Grass Valley-Nevada City, and Alleghany districts.

❑ A placer deposit is a concentration of natural material that has accumulated in the unconsolidated sediments of a stream bed. Heaviness and resistance to corrosion make gold an ideal substance to accumulate in placer deposits. Panning is the simplest method of separating the gold from the silt, sand, and gravel of the stream deposits. It is the method most commonly used by the beginning prospector.

❑ The gold pan, an indispensable prospecting tool, is versatile, efficient, inexpensive, and portable. With some practice, even an inexperienced panner should be able to recover most of the gold from the sand and gravel. Gold is much heavier than the minerals that make up the sand and gravel; when the water is shaken and swirled, the grains and flakes of gold sink and collect on the bottom of the pan. As the lighter materials are winnowed from the gold, they are removed from the pan.

❑ Many placer districts in California were mined on a large scale in the mid-1950s and as recently as the mid-1980s. The streams that drain the rich Mother Lode—the Feather, Mokelumne, Sacramento, American, Cosumnes, Calaveras, Yuba, and Trinity rivers in northern California, have concentrated considerable gold in gravels.

A REMINDER!

You must always obtain permission from the owner to prospect on private property. Before prospecting on federal land, contact the U.S. Bureau of Land Management at (916) 978-4400.

A recommended source of information is U.S. Bureau of Land Management Special Publication 1996, *Location and Patenting of Mining Claims and Mill Sites in California.*

CALIFORNIA GEOLOGICAL SURVEY
CALIFORNIA CONSERVATION

Revised 4/2002

PAGE 2

Offices of the California Geological Survey

Publications and Information Office
801 K Street, MS 14-33
Sacramento, CA 95814-3532
(916) 445-5716

Southern California Regional Office
655 South Hope Street, Suite 700
Los Angeles, CA 90017-3231
(213) 239-0878

Bay Area Regional Office
185 Berry Street, Suite 210
San Francisco, CA 94107-1728
(415) 904-7707

Selected Gold-Related Publications of the California Geological Survey

- CD98-001 *California Gold Mines: A Sesquicentennial Photograph Collection on CD-ROM.* 1998
- B193 *Gold Districts of California, Sesquicentennial Edition: California Gold Discovery to Statehood* (includes new mine list). Scale 1:1,000,000. 1998
- MM009 *Map of California Historic Gold Mines. The California Gold Discovery to Statehood Sesquicentennial (1998-2000) Edition.* Scale 1:1,500,000. 1998
- *The Elephant as They Saw It: A Collection of Contemporary Pictures and Statements on Gold Mining in California (1500-1860).* 1997
- SP098 *Fluvial Geomorphology and River-Gravel Mining: A Guide for Planners, Case Studies Included.* 1990
- SP103 *Mines and Mineral Producers Active in California (1997-98).* Scale 1:500,000. 1999
- SP034 *Geology of Placer Deposits.* 1970.
- SP041 *Basic Placer Mining.* 1946
- SP087 *Placer Gold Recovery Methods.* 1986

To order one of these publications (or any other California Geological Survey publication) call 916-445-6199.

Graphics by Dinah D. Maldonado, California Geological Survey.

Visit the California State Mining and Mineral Museum in Mariposa, along historic Highway 49.
P.O. Box 1192
Mariposa, CA 95338
(209) 742-7625

www.conservation.ca.gov

THE RESOURCES AGENCY
MARY NICHOLS
SECRETARY FOR RESOURCES

STATE OF CALIFORNIA
GRAY DAVIS
GOVERNOR

DEPARTMENT OF CONSERVATION
DARRYL YOUNG
DIRECTOR

Geologic Map of Julian

Source: Todd, Victoria R., U.S.G.S., Geologic Map of the Julian 7.5' Quadrangle, San Diego County, California, 2015.

LIST OF MAP UNITS

Qya **Younger alluvium (Quaternary)**—Sand, silt, and gravel in modern washes

Qc **Colluvium (Quaternary)**—Sand, silt, and gravel on hillslopes; grades locally to younger alluvium

Qoa **Older alluvium, undivided (Quaternary)**—Dissected sand and gravel of alluvial fans and terraces in modern washes.

Qoa_1 **Older alluvium (Quaternary)**—Older materials of alluvial fan on west side of San Felipe Valley

Qoa_2 **Older alluvium (Quaternary)**—Younger materials of alluvial fan on west side of San Felipe Valley

Qoa_3 **Older alluvium (Quaternary)**—Coarse alluvium beneath slide or thrust of crystalline rocks in the east wall of Banner Canyon

Qoa_4 **Older alluvium (Quaternary)**—Alluvium at southeast end of Rodriguez Canyon composed of boulders of tonalite of Granite Mountain

Qls **Landslide deposits (Quaternary)**—Landslides, slumps, and debris flows; arrows indicate direction of movement

Kl **Leucocratic dikes (Cretaceous)**—Dikes consisting of pegmatite, alaskite, aplite, and leucogranite

Kgm **Tonalite of Granite Mountain (Late Cretaceous)**—biotite-hornblende tonalite; hornblende-biotite tonalite and lesser granodiorite; minor quartz diorite. Medium- to coarse-grained; weak to mylonitic foliation. C.I. from 17-27

Kc **Cuyamaca Gabbro (Early Cretaceous)**—Interior of large plutons: hornblende-bearing troctolite; anorthositic gabbro ± amphibole ± opx ± olivine; amphibole-olivine gabbronorite; minor hornblende diorite, leucodiorite ± pyroxene ± biotite. Fine- to medium-grained smaller bodies, marginal zones of large plutons: hornblende gabbro ± opx ± cpx ± biotite. Moderately to strongly foliated

KJem **Quartz diorite of East Mesa (Cretaceous and Jurassic)**—Fine- to medium-grained, gneissic biotite-hornblende tonalite and quartz diorite; and fine-grained (quenched), locally porphyritic biotite-hornblende quartz diorite and tonalite; lesser diorite, granodiorite, quartz monzodiorite, and gabbro. Texturally and compositionally heterogeneous. Strongly foliated to mylonitic. Some rocks contain hypersthene ± clinopyroxene

Jcr **Granodiorite of Cuyamaca Reservoir (Jurassic)**—Biotite- and hypersthene-biotite granodiorite and tonalite ±actinolitic amphibole ± hornblende. Contains sphene, ilmenite, allanite. Fine- to medium-grained, gneissic, locally mylonitic. C.I. = 16-30

Jhc **Harper Creek Gneiss (Jurassic)**—Biotite granodiorite and tonalite, lesser monzogranite; mylonitic rocks are quartz-rich granitoids. Contains graphite, muscovite, tourmaline, ilmenite, sillimanite (mostly fibrous habit), corderite, garnet. Fine- to coarse-grained, gneissic to mylonitic. Average C.I. is 22

JTrm **Julian Schist (Jurassic and Triassic)**—Interlayered, intergradational semi-pelitic, pelitic, and quartzitic schists; calcsilicate-bearing feldspathic metaquartzite; and minor small-pebble metaconglomerate, metatuff, and orthoamphibolite; metamorphosed to amphibolite facies. Protoliths interpreted as turbidites deposited on submarine fan complex that received intermittent basaltic flows and sills

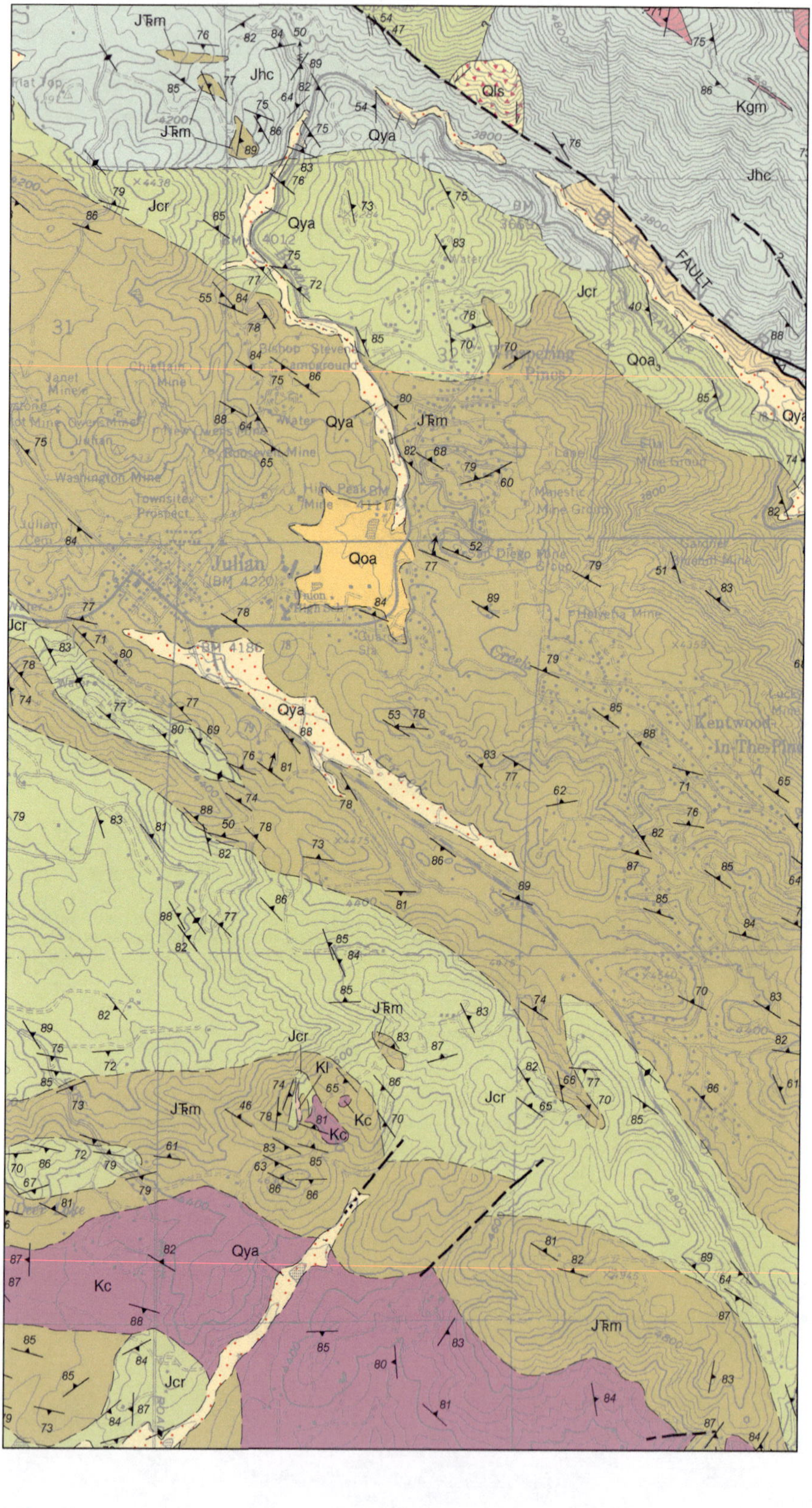

JTRm
Jhc
Qya
Qls
Kgm
Jcr
Qoa
FAULT
Bishop Stevens Campground
Chieftain Mine
Janet Mine
Owens Mine
Washington Mine
Townsite Prospect
High Peak Mine
Julian
Julian Cem
Whispering Pines
Ella Mine Group
Majestic Mine Group
San Diego Mine Group
Helvetia Mine
Kentwood-In-The-Pines
Union High Sch
Kl
Kc
BM 4012
BM 4186
31
32

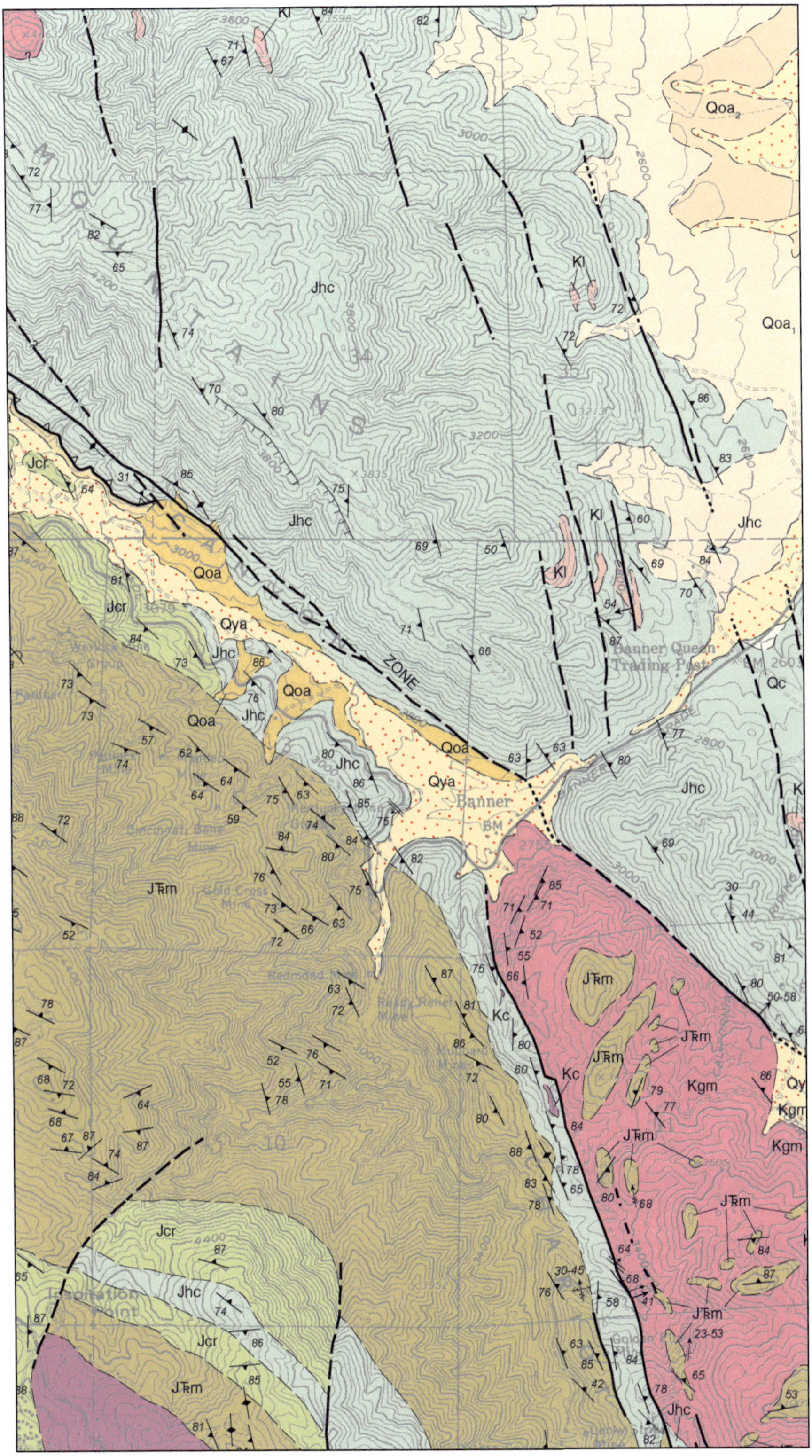

MOUNTAINS
ZONE
Banner
Banner Queen Trading Post
Inspiration Point
Qoa$_2$
Qoa$_1$
Qoa
Qya
Qc
Jhc
Jcr
JTRm
Kl
Kc
Kgm

CHRISTOPHER D. LIVESEY was born in 1981 and has lived throughout southern California. A Bachelor of Science Degree focusing on geology was obtained at the California State University of Long Beach. During and after undergraduate study, Mr. Livesey's career was focused in consulting engineering geology. During his consulting career, several licenses and certifications accredit his accomplishments, which have included the title as recognized by the State of California as a Professional Geologist, holding a City of Los Angeles Deputy Grading Inspection license, certification through the American Concrete Institute as a Grade 1 concrete tester, several California State Department of Transportation testing certificates, as well as other miscellaneous certifications. While robust accreditations complimented Mr. Livesesy's career skill set, outside career activities include volunteer service to the San Diego Association of Geologists throughout a total of four years, which included positions as treasurer, secretary, vice president and (in 2018) president. General point of contact: julianbannerminingdistrict@gmail.com